中国轻工业"十三五"规划教材

高等学校食品质量与安全专业适用教材

食品质量与安全专业英语

王颖　蓝蔚青　主编

中国轻工业出版社

图书在版编目（CIP）数据

食品质量与安全专业英语/王颖，蓝蔚青主编. ——
北京：中国轻工业出版社，2023.10
ISBN 978-7-5184-3337-7

Ⅰ.①食… Ⅱ.①王… ②蓝… Ⅲ.①食品安全—质量管理—英语—教材 Ⅳ.①TS201.6

中国版本图书馆CIP数据核字（2020）第259473号

责任编辑：马　妍　　责任终审：许春英　　封面设计：锋尚设计
文字编辑：巩孟悦　　责任校对：吴大朋　　责任监印：张　可
策划编辑：马　妍　　版式设计：砚祥志远

出版发行：中国轻工业出版社（北京东长安街6号，邮编：100740）
印　　刷：河北鑫兆源印刷有限公司
经　　销：各地新华书店
版　　次：2023年10月第1版第1次印刷
开　　本：787×1092　1/16　印张：10.5
字　　数：242千字
书　　号：ISBN 978-7-5184-3337-7　定价：48.00元
邮购电话：010-65241695
发行电话：010-85119835　传真：85113293
网　　址：http://www.chlip.com.cn
Email：club@chlip.com.cn
如发现图书残缺请与我社邮购联系调换
181312J1X101ZBW

本书编写人员

主　　编　王　颖　黑龙江八一农垦大学
　　　　　蓝蔚青　上海海洋大学
副 主 编　王淑梅　哈尔滨学院
　　　　　何胜华　许昌学院
　　　　　范秀萍　广东海洋大学
参编人员　（按姓氏笔画排列）
　　　　　卫晓怡　上海商学院
　　　　　卢　瑛　上海海洋大学
　　　　　刘书成　广东海洋大学
　　　　　刘俊梅　吉林农业大学
　　　　　许　倩　塔里木大学
　　　　　孙晓红　上海海洋大学
　　　　　李振兴　中国海洋大学
　　　　　佟立涛　中国农业科学院农产品加工研究所

前言 Preface

食品质量与安全专业英语是食品类相关专业的一门专业必修课程,读者通过本课程的学习,可以获得食品质量与安全、食品科学与工程等相关方面的英语知识和英语使用能力,了解食品质量安全等方面中外用语间的相互联系,并为今后从事食品科学与工程等相关领域的技术工作打下基础。

国内大部分院校自教育改革以来,对专业基础课程在学时上进行了一定程度的压缩,同时,教学理念和授课对象均发生了较大变化。在这种新形势下,作者在精简内容的同时扩大了知识涵盖面,使学生在有限的学时内拓展知识面并加深理解。全书分别从食品安全控制、食品原料成分和营养、食品添加剂、食品毒理学、食品包装、食品工厂卫生、食品安全管理体系、食品质量安全国际组织、食品质量安全认证等方面进行介绍。

本书附所有课文的中文翻译,常用词、词组,以及常用缩写词。每课课后还附有练习题,可供读者巩固知识,检查学习效果。另外,本书融入数字资源,读者可通过扫描书中二维码获取音频资源,与教材配合使用,加深章节内容的理解与认识,并纠正英语发音。

本书共十课,编写分工如下:第一课由王淑梅、佟立涛承担,第二课由王淑梅承担,第三课由何胜华承担,第四课由蓝蔚青、卢瑛承担,第五课由孙晓红承担,第六课由王颖承担,第七课由许倩承担,第八课由范秀萍、刘书成和卫晓怡承担,第九课由李振兴承担,第十课由刘俊梅承担。全书由黑龙江八一农垦大学文理学院孟庆玲,爱丁堡大学张乐,利兹大学孙泽堃审校完成。邀请利兹大学食品科学与营养学院 Jackson Sun 担任英文音频的录音。在此一并表示感谢。

本书介绍内容深入浅出,难易适度,适用性强,学术性与普及性兼顾,主要适用于高等院校食品科学与工程、食品质量与安全及相关专业学生的必修课、选修课教学或作为专业自学考试用教材。

由于编者水平有限,编写时间仓促,书中难免有错误或不妥之处,敬请读者批评指正。

编者
2023.3

| 目录 | Contents

LESSON 1　OVERVIEW OF FOOD SAFETY ······························· 1
　　Vocabulary ··· 4
　　Exercises ··· 5
　　参考译文 ··· 6

LESSON 2　FOOD RAW MATERIAL INGREDIENTS AND NUTRITION ············ 9
　　Vocabulary ·· 16
　　Exercises ·· 17
　　参考译文 ·· 18

LESSON 3　ISSUES IN FOOD SAFETY ···································· 25
　　Vocabulary ·· 34
　　Exercises ·· 35
　　参考译文 ·· 37

LESSON 4　FOOD PRESERVATIVES ······································ 45
　　Vocabulary ·· 49
　　Exercises ·· 49
　　参考译文 ·· 51

LESSON 5　FOOD RISK ANALYSIS AND FOOD QUALITY ··················· 54
　　Vocabulary ·· 57
　　Exercises ·· 58
　　参考译文 ·· 59

LESSON 6　FOOD SAFETY TRACING ···································· 62
　　Vocabulary ·· 68
　　Exercises ·· 70
　　参考译文 ·· 71

LESSON 7   FOOD SAFETY AND MICROBIAL INFLUENCE FACTORS ············ 77
   Vocabulary ··· 81
   Exercises ··· 84
   参考译文 ··· 86

LESSON 8   FOOD QUALITY MANAGEMENT ································ 90
   Vocabulary ··· 97
   Exercises ··· 98
   参考译文 ··· 100

LESSON 9   QUALITY CONTROL TOOLS AND FOOD SAFETY EVALUATION
   ·· 106
   Vocabulary ··· 118
   Exercises ··· 119
   参考译文 ··· 121

LESSON 10   CRISIS MANAGEMENT FOR FOOD INDUSTRY ············ 131
   Vocabulary ··· 136
   Exercises ··· 137
   参考译文 ··· 138

References ··· 143

练习题答案 ··· 148

LESSON 1

OVERVIEW OF FOOD SAFETY

1 INTRODUCTION

Illness and death from diseases caused by contaminated food are constant threats to public health. The increasing negative effects on food safety have put more people at risk of carcinogenic diseases. In developed countries, consumer demographics have changed, resulting in an increased number of consumers who are more susceptible to foodborne disease (FBD). In developing countries, contaminated water contributes to the contamination of food during production and harvest as well as during final food preparation activities. Nowhere has that situation been more complex and challenging than in China, where a combination of pollution and an increasing food safety risk has affected a large part of the population.

The most frequent causes of FBD are diarrheal disease agents, particularly *Campylobacter* spp. In addition, agents such as non-typhoidal *Salmonella enterica*, are also responsible for the majority of deaths due to FBD in all regions of the world.

Another cause of FBD is mycotoxins. Mycotoxins are toxic secondary metabolites of fungal origin and contaminate agricultural commodities before or under post-harvest conditions. When ingested or absorbed through the digestive tract, mycotoxins will cause sickness or death of humans and animals.

This chapter mainly introduces some impact factors on food safety, including microbiological activities, environment contaminants and some animal production.

2 MICROBIOLOGICAL ACTIVITIES

There are many diseases resulting from the consumption of food containing pathogenic microorganisms or their toxic metabolites. Because microorganisms are abundant in the field and water where foods are produced and harvested, they are commonly associated with the finished products.

2.1 Bacteria and Viruses

There is a wide variety of bacteria that cause foodborne infections. Non-typhoidal *Salmonella* and

Campylobacter are believed to be the most prevalent in developed countries, estimated on rates of Yersinia infections. While the major foodborne viruses of significance are caliciviruses and hepatitis viruses.

2.2 Mycotoxins

It is reviewed that mycotoxin contamination in agricultural commodities causes a variety of diseases in humans and animals. Foods susceptible to mycotoxin contamination and of concern globally include groundnuts, maize, rice, and sorghum.

Exposure of humans to mycotoxins varies by geographical location due to the differences in agricultural crops farming practices and climate. Therefore, diseases resulting from ingestion of mycotoxins range from acute conditions (mycotoxicosis) to chronic conditions (cancer).

3 ENVIRONMENT CONTAMINANTS

Pesticide application and chemical pollutants are considered to be the most important enviromental factors impacting food safety in China in recent years. During the growth of crops or livestock, they may be exposed to hazardous environmental contaminants present in the local region in China.

3.1 Persistent Organic Pollutants (POPs)

POPs are a class of organic compounds that are produced industrially for use as insecticides and as plasticizers in a variety of products. Long-term exposure to POPs has been linked to reproductive disorders, immune system dysfunction, nerve damage and even cancers. POPs have been found in measurable levels in a variety of animals with contaminated blood and breast milk. Pesticide residues are most often found in fruits, grains, and vegetables. It is estimated that three million people suffer from pesticide poisoning annually in the world.

3.2 Heavy Metals

Heavy metal emissions from processing facilities enter the environment and contaminate air and soil which can lead to the contamination of drinking water and food crops. In some regions in China, heavy metals have caused serious agricultural land and food pollution. The major heavy metals of concern in the food pollution are lead, cadmium, mercury and arsenic. Additionally, chromium, selenium, tin, antimony, copper, thallium, fluoride and zinc also pose potential health threats to consumers.

4 HEALTH HAZARDS INHERENT TO FOODS

Generally speaking, natural toxins and allergens present in foods may lead to adverse health conditions. Plant toxins are particularly problematic in livestock production due to the presence of poisonous plants growing in grazing areas throughout the world. Over 170 different foods have been identified to potentiate an allergic response. Symptoms, as a result of contact with allergens, vary from minor to severe.

5 ANIMAL PRODUCTION

In recent years, a number of antibiotics have been used in agricultural livestock industries for therapeutic, prophylactic, and growth promotion reasons. Antibiotic residues in animal products may impact food safety and threatens human health directly or indirectly. It is estimated that 8~12 million kg of antibiotics are used in livestock production in the United States annually. In 2015, about 180 million kg of raw antibiotic ingredients were used in agriculture to prevent disease and improve production, with up to 70% released directly or indirectly into the environment.

6 FOOD SAFETY OVERVIEW

Food safety is now a commonplace concept. But it's also a relatively young field and continues to change rapidly. The World Food Summit in 1996 defined food security as a situation in which "all people, at all times, have access to sufficient, safe and nutritious food to meet their dietary needs and food preferences for an active and healthy life".

The book is aimed to introduce the main background and framework of the topic, review the issues of foodborne disease and pay special attention to emergent and resistant pathogens. These include bacterial, fungal, and viral agents. Some quality management tools used in food industry, the variety and quantity of food additives, important quality management tools, such as Good Manufacturing Practices (GMP), Sanitation Standard Operating Procedures (SSOP), and Hazard Analysis and Critical Control Point (HACCP) are introduced. The main objective of the HACCP system is prevention of risks to human health, as well as prevention of changes in foodstuffs by means of control practices in production steps in which there is a greater probability of health hazards. HACCP may be applied to all production steps, from raw material production to the manufacture of the final product. Before the HACCP system is implemented in food industries, prerequisite programs, such as GMP and SSOP have to be established. Some comprehensive descriptions of the GMP, SSOP, and HACCP principles and practices are presented, taking into account their positive impact as important safety strategies in the food industry and the barriers to its implementation. The main food risk analysis, food quality tracing, food quality management, and food safety evaluation are reviewed. Keeping in mind consumer demand for more natural and safe foods, the book includes some significant contents that are focused on the most foodborne infections, which can cause consumer foodborne illness. Thus, the standards of identities of specific food products and government are important measures for food quality. Therefore, in order to ensure that food safety education programs for consumers are effective in achieving their goals and preventing foodborne illness, it is important that these programs incorporate a rigorous and systematic program evaluation. At last, crisis management for food industry is also reviewed, and crisis management innovation should be put in an important position in Chinese food enterprises. Food industry is a moral industry. The crisis management of food enterprises must be emphasized and put in the leading position of enterprises. Food enterprises should also constantly seek technological innovation, market management and organization system.

Lesson 1

Vocabulary

/ Words /

carcinogenic [ˌkɑːrsɪnəˈdʒenɪk] adj. 致癌的
demographics [ˌdeməˈɡræfɪks] n. 人口统计资料
majority [məˈdʒɔːrəti] n. 大多数
typhoid [ˈtaɪfɔɪd] adj. 伤寒的
maize [meɪz] n. 玉米
chronic [ˈkrɑːnɪk] adj. 慢性的
prophylactic [ˌproʊfəˈlæktɪk] adj. 预防性的
emissions [ɪˈmɪʃnz] n. 排放物
symptoms [ˈsɪmptəmz] n. 症状

contamination [kənˌtæmɪˈneɪʃn] n. 污染
susceptible [səˈseptəbl] adj. 易得病的
pathogenic [ˌpæθəˈdʒenɪk] adj. 致病的
prevalent [ˈprevələnt] adj. 盛行的
sorghum [ˈsɔːrɡəm] n. 高粱
pesticide [ˈpestɪsaɪd] n. 杀虫剂
plasticizer [ˈplæstəˌsaɪzər] n. 塑化剂
allergen [ˈælərdʒən] n. 过敏原
therapeutic [ˌθerəˈpjuːtɪk] adj. 治疗的

/ Phrases /

Campylobacter spp.　　弯曲杆菌
Salmonella enterica　　沙门氏菌

/ Abbreviations /

foodborne disease（FBD）食源性疾病
Persistent organic pollutant（POP）持久性有机污染物
Good Manufacturing Practices（GMP）良好生产规范
Sanitation Standard Operating Procedures（SSOP）卫生标准操作程序
Hazard Analysis and Critical Control Point（HACCP）危害分析和关键控制点

Exercises

I. Write true or false for the following statements according to the passage

1. (　) There is a wide variety of bacteria that cause foodborne infections.
2. (　) Foods susceptible to mycotoxin contamination and of concern globally include groundnuts, milk, meat, and egg.
3. (　) Exposure of humans to mycotoxins varies by geographical location due to the differences in agricultural crops farming practices and climate.
4. (　) Pesticide application and chemical pollutants are considered to be the most important enviromental factors impacting on food safety in China recent years.
5. (　) POPs are a class of inorganic compounds.
6. (　) The major heavy metals of concern in the food pollution are iron, copper, manganese, and arsenic.
7. (　) Generally speaking, natural toxins and allergens present in foods may lead to adverse health conditions.
8. (　) Antibiotic residues in animal products may impact food safety and threatens human health directly or indirectly.

II. Answer the following questions according to the passage

1. What causes foodborne disease?
2. What are the hazards of persistent organic pollutants?
3. What are the main heavy metal contamination in food?

III. Fill in the blanks according to the passage

1. Foods susceptible to mycotoxin contamination and of concern globally include groundnuts, _____, _____, and _____.
2. The major heavy metals of concern in the food pollution are _____, _____, _____, and arsenic.

IV. Translate the following words and expressions into Chinese

foodborne disease (FBD)　　　　　　post-harvest
microbiological activity　　　　　　　micro-organisms
finished products　　　　　　　　　environment contamination
persistent organic pollutants (POPs)　organic compounds
food additives　　　　　　　　　　antibiotic

V. Translate the following expressions into English

1. 加工企业重金属排放进入环境，污染空气和土壤，从而导致饮用水和粮食作物的污染。在中国的一些地区，重金属已经造成了严重的农业用地和食品污染。
2. 食用含有致病微生物或其有毒代谢物的食物会引发多种疾病。因为在食品生产和收获的环境中或水里都存在大量微生物，因此微生物通常存在于成品中。

参考译文

第1课 食品安全概述

1 简介

受污染食物引起的疾病和死亡给公众健康带来持续性威胁。日益增加的食品安全负面影响使更多的人遭受致癌的风险。在发达国家，消费人群结构的变化增加了对食源性疾病（FBD）的易感性。而在发展中国家，污染的水使粮食在生产和收获期间以及在最后的加工过程发生污染。在中国，这种情况更为复杂和具有挑战性，污染和日益增加的食品安全风险已经影响了大部分人口。

FBD 最常见的病因是腹泻病原体，尤其是弯曲杆菌；另外如非伤寒沙门氏菌，世界各地 FBD 引发的大多数死亡都与之有关。

FBD 的另一个病因是真菌毒素。真菌毒素是真菌来源的有毒次生代谢产物，在收获前或在收获后的条件下污染农产品。当经消化道吸收时，毒素会引发人类和动物的疾病或死亡。

本章主要介绍了影响食品安全的一些因素，包括微生物活性、环境污染物和一些动物生产。

2 微生物活性

食用含有致病微生物或其有毒代谢物的食物会引发多种疾病。因为在食品生产和收获的环境中或水里都存在大量微生物，因此微生物通常存在于成品中。

2.1 细菌和病毒

有多种细菌可引起食源性感染。据报道，发达国家耶尔森氏菌感染最常见的是由非伤寒沙门氏菌和弯曲杆菌引发的。主要的食源性病毒是杯状病毒和肝炎病毒。

2.2 真菌毒素

据报道，农产品中真菌毒素可引发人类和动物的多种疾病。受真菌毒素污染的全球性食物包括花生、玉米、大米和高粱。

由于农作物种植方式和气候的差异，人类受真菌毒素的污染因地域不同而不同。因此，由真菌毒素引发的疾病有急性（真菌中毒），也有慢性（癌症）。

3　环境污染物

农药的使用和化学药物污染是近年来影响我国食品安全的最重要因素。在农作物或家畜生长期间，都可能接触当地的有害环境污染物。

3.1　持久性有机污染物

有机污染物是一类有机化合物，在工业上用作杀虫剂，在各种产品中用作增塑剂。长期接触有机污染物与生殖障碍、免疫系统功能障碍、神经损伤甚至癌症有关。在许多动物的血液和乳汁中都已检测出这类污染物。水果、谷物和蔬菜的农药残留最常见。据估计，全世界每年有300万人农药中毒。

3.2　重金属

加工企业重金属排放进入环境，污染空气和土壤，从而导致饮用水和粮食作物的污染。在中国的一些地区，重金属已经造成了严重的农业用地和食品污染。食品污染主要的重金属是铅、镉、汞和砷。此外，铬、硒、锡、锑、铜、铊、氟化物和锌也对消费者构成潜在的健康威胁。

4　食品固有的健康危害

一般来说，食物中的天然毒素和过敏原可能会导致健康状况的恶化。植物毒素是牲畜生长过程的主要威胁，因为世界各地的牧区都生长着有毒植物。有超过170种不同的食物被鉴定出可能引起过敏反应。接触过敏原后引发的过敏症状从轻微到严重不等。

5　动物生产

近年来，一些抗生素被用于农业、畜牧业中，用于治疗、预防疾病和促进动物生长。动物产品抗生素的残留会影响食品的安全，对人体健康造成直接或间接的危害。据估计，美国每年有800万~1200万kg的抗生素用于牲畜生产。仅2015年，大约1.8亿kg抗生素原料产品用于农业来预防疾病和提高产量，而其中70%直接或间接地被排放到环境中。

6　食品安全概述

食品安全是当下的一个普遍概念。但它也是一个相对崭新的领域，且不断快速发展。1996年，世界粮食首脑会议将粮食安全定义为"所有人在任何时候都能获得充足、安全和有营养的食物，以满足其积极健康生活的饮食需要和食物偏好"的状况。

本书目的是介绍食品安全的主要背景和框架，概述食源性疾病等问题，并着重论述突发性和耐药性病原体。这些病原体包括细菌、真菌和病毒。同时也介绍了食品工业中的一些质量管理工具和食品添加剂的种类和数量，包括重要的食品质量管理工具如良好操作规范（GMP）、卫生标准操作规程（SSOP）和危害分析与关键控制点（HACCP）等。HACCP体系主要目的是预防引发人类健康的风险，通过控制生产加工中存在较大安全危害的步骤来防止食品质量的改变。HACCP体系可应用于从原材料生产到最终产品的所有生产过程。在食品行业实施HACCP体系之前，必须建立必要的程序，如GMP和SSOP。本书对GMP、SSOP和HACCP的原理和实际应用进行了全面描述，既考虑以上体系作为食品工业重要安全保障的积极影响，也考虑其实施过程的障碍。本书对食品风险分析、食品质量追溯、食品质量管理和食品安全评价进行了综述。考虑到消费者对更天然和更安全的食品需求，本书也涵盖了一些食源性感染等重要内容，这些食源性感染会引发消费者食源性的疾病。因此，特定食品的标识标准和政府是衡量食品质量的重要保障。因此，为确保消费者的食品安全教育培训有效落实，并能预防食源性疾病，将这些培训纳入严格的、系统的项目评估很重要。最后，本书还对食品行业的危机管理进行了论述，指出危机管理创新应放在中国食品企业的重要地位。食品工业是一项道德工业。食品企业的危机管理必须得到重视，并置于企业的主导地位。食品企业也应不断寻求技术创新、市场管理和组织体制。

LESSON 2

FOOD RAW MATERIAL INGREDIENTS AND NUTRITION

1 INTRODUCTION

Ingredients in foods can be divided into two main categories, namely, active and inactive. Active ingredients can be considered as those that supply energy to the body or serve as its nutrient foods and some food additives. Inactive ingredients do not exert physiological actions when ingested or applied to the body. Their primary function is to act as diluents or to facilitate the ultimate intake or utilization of the active ingredients, such as dietary fiber. This chapter is an introduction of food raw materials with emphasis on food raw material ingredients and nutrition.

2 MEAT

2.1 Introduction

Meat is an increasingly important source of high-value animal protein worldwide. Meat and meat products, are referred to here as red meats, or postmortem muscles from mammalian species. Beef, veal, pork, and lamb/mutton are an important component in diet.

Meat and meat products are considered as sources of cholesterol in the diet. Therefore, in most developed countries, high meat intake contributes to higher total saturated fatty acid (SFA) intake than recommended. Therefore, meat and meat products are frequently criticized by nutritionists and national health authorities, who advise consumers to choose lean meats and low-fat meat products.

2.2 Meat and Nutrition

Meat consists of several tissues, such as muscle fibers, connective and adipose tissues. Muscle fibers, connective tissue, and fat play key roles in the determination of meat and fish flesh quality. Generally speaking, skeletal muscle consists of approximately 90% muscle fibers and of 10% connective and other components. As a result, a meat that is rich in proteins with a high proportion of essential amino acids and polyunsaturated fatty acids is considered to exhibit good nutritional quality.

Meat fat comprises mostly monounsaturated and saturated fatty acids, with oleic (C18∶1), palmitic (C16∶0), and stearic acid (C18∶0) being the most ubiquitous. Nutritional value is affected by the balance between saturated fatty acids (SFA) and polyunsaturated fatty acids (PUFA). In all species, the bulk of PUFA are deposited in phospholipid (PL) in muscle. Meat is also a significant source of long-chain $n-3$ fatty acids, the intake of which is widely judged to be too low at present. However, high levels of PUFA in meat can lead to lipid oxidation postmortem which may negatively affect the flavor of cooked meat and the color of displayed meat.

Despite the surge in poultry product consumption in the past two decades, red meats remain to have a dominant market share in all muscle foods produced in the world. Many cured meat products are available in the current food market, for example, traditional pork products (ham and bacon), beef products (corned beef).

3 SEA FOODS

3.1 Introduction

The increasing demand for seafood may be due to its health benefits, especially, fish and shellfish. In recent years, various fish and shellfish source materials such as skin, muscle, frame, bone, and internal organs are presently utilized to isolate a number of bioactive materials. Skin and protein remaining on fish frame could be used as cheap materials for identification and isolation of bioactive peptides. Fish bone is a good source of calcium.

3.2 Fish and Nutrition

Collagen and gelatin are the major structural proteins found in the skin and bones of fish. Collagen is commonly used in medical and pharmaceutical industries especially as drug carriers, based on structural roles and better compatibility within the human body. Gelatin is a heterogeneous mixture of high molecular-weight and water-soluble proteins, and it has a unique amino acid arrangement in its sequence and contains relatively high amounts of glycine, alanine, and proline.

Fish frames resulting from filleting operations contain a considerable amount of muscle protein. Frames of some fish species, an ideal protein source, have been utilized to obtain functional hydrolysates.

Generally speaking, approximately 65%~70% of fish bones is composed of inorganic substances. Almost all of these inorganic substances are hydroxyapatite composed of calcium, phosphorus, oxygen, and hydrogen.

3.3 Shellfish and Nutrition

Shellfish provide high-quality proteins with all the dietary essential amino acids for maintenance and growth of the human body. Shellfish also provide saturated, monounsaturated, and polyunsaturated fats to a healthful diet. Previous research suggests that calcium compounds obtained from oyster shell and fish gelatin-derived peptides combined could be used as an effective dietary calcium source.

4 EGG

4.1 Introduction

Eggs are now considered a highly nutritious food with unique components and the referential protein worldwide by the World Health Organization (WHO). Eggs have numerous functions in food production, and these functions make them the preferred ingredients in many food formulations. However, the egg has been subjected to negative publicity generally related to the cholesterol content for many years.

Eggs are considered nature's most perfect food containing an excellent source of protein of high biological value, high ratio of unsaturated fatty acids to saturated fatty acids, and an excellent source of minerals and all the vitamins.

4.2 Components in Egg and Nutritional Value

An average hen egg weighs about 57 g, which includes the weight of the yolk, white, and shell. Each component differs in composition, as shown in Table 1.

Table 1   Chemical composition of the hen's egg by percentage

Components	Total/%	Water/%	Protein/%	Fat/%	Ash/%
Whole egg	100	65.5	11.8	11.0	11.7
Egg white	58	88.0	11.0	0.2	0.8
Egg yolk	31	48.0	17.5	32.5	2.0
Shell	11	—	—	—	—

Source: USDA

4.2.1 The Yolk

An egg yolk comprises approximately 30%~33% of the total egg weight. Egg yolk provides all of the egg's cholesterol, almost all of the fat, half of the protein, most of the calcium, phosphorus, iron, zinc, vitamins B6, B12, and A, folic acid, half of the riboflavin and thiamine.

Therefore, generally, the yolk has a higher nutrient density than the white, containing all of the vitamins known except vitamin C. The yolk contains triglycerides, phospholipids, and sterols. The primary phospholipid is phosphatidyl choline or lecithin. Protein in the yolk is primarily vitellin which is present in a lipoprotein complex as lipovitellin and lipovitellinin.

4.2.2 Egg White

Egg white contains about half of the protein and riboflavin. The white of the egg represents approximately 55%~60% of the total egg weight.

4.2.3 Air Cell, Membranes, and Shell

The air cell is developed by a separation of the two shell membranes, usually at the large end of the egg, as the egg contents shrink during cooling. The next layers of the egg are the inner and outer shell membranes. These relatively thin keratin-like membranes are one of the egg's chief defenses against bacterial invasion.

The shell comprises 9%~12% of the total egg weight. It consists largely of calcium carbonate (94%) with some magnesium carbonate (1%), calcium phosphate (1%), and organic matter, chiefly protein (4%).

5 MILK AND MILK PRODUCTS

5.1 Introduction

Milk and milk products have long traditions in human nutrition and contain numerous essential nutrients. Human beings consume large amounts of milk of a few species, such as cows, water buffaloes, goats, and sheep, especially cows.

Milk consists of water, lipids, carbohydrates, proteins, salts, and a long list of miscellaneous constituents. Milk contains the nutrients needed for growth and the development of the youth, and is a resource of lipids, proteins, amino acids, vitamins, and minerals. Also it contains immunoglobulins, hormones, growth factors, cytokines, nucleotides, peptides, polyamines, enzymes, and other bioactive peptides.

5.2 Components in Milk and Nutritional Value

5.2.1 Protein

Bovine milk contains about 3.3 g protein/100g (Table 2). The milk protein has a high biological value, and milk is therefore a good source of essential amino acids. In addition, milk contains a wide range of proteins with biological activities ranging from antimicrobial ones to those facilitating absorption of nutrients, as well as acting as growth factors, hormones, enzymes, and antibodies.

Table 2　　　　　　　　Basic composition of goat, cow, and human milks

Constituents	Goat	Cow	Human
Fat/ (g/100g)	3.8	3.6	4.0
Protein / (g/100g)	3.5	3.3	1.2

Continued table

Constituents	Goat	Cow	Human
Lactose / (g/100g)	4.1	4.6	6.9
Total Solids / (g/100g)	12.2	12.3	12.3
Calories/ (cal/100g)	70	69	68

The nitrogen in milk is distributed among caseins, whey proteins, and non-protein nitrogen. Milk protein content has been determined by the analysis for nitrogen (N). This has the advantage that nitrogen is a major constituent, comprising about one-sixth of the mass of the protein. Nitrogen contents of the individual milk proteins are nearly the same. Multiplication by 6.25 has been used commonly to convert nitrogen content to protein on the basis of 16% nitrogen in milk proteins.

The casein content of milk represents about 80% of milk proteins and almost all of the caseins are associated with calcium and phosphate. Therefore, casein's biological function is to carry calcium and phosphate.

The other milk proteins, called whey proteins, are a diverse group. Whey proteins are the liquid remaining after milk has been curdled to produce cheese, and are used in many products for human consumption. The milk whey proteins are globular proteins that are more water soluble than caseins, and the principle fractions are beta-lactoglobulin, alpha-lactalbumin, bovine serum albumin and immunoglobulins. Whey proteins are considered as rapidly digested protein that gives high concentrations of amino acids in postprandial plasma.

5.2.2 Lipids

On average, cow milk contains about 3.6g total lipid/100g, and goat milk is 3.8. Milk is in the form of a stable emulsion of fat globules dispersed throughout an aqueous phase containing the non-fat solids. Smaller fat globules would be better for digestion action. Average diameters of fat globules for goat, cow, buffalo, and sheep milks were reported to be 3.5, 4.5, 5.9, 3.3 μm, respectively.

Triacylglycerols, which account for about 95% of the lipid fraction, are composed of fatty acids of different length (4~24 C-atoms) and saturation. A diet rich in milk fat therefore may help to increase this ratio of the total dietary fatty acids.

More than half of milk fatty acids are saturated. The saturated fatty acids, consisting of lauric acid, myristic acid (14 : 0) and palmitic acid (16 : 0), raise blood cholesterol levels. Diets rich in saturated fat have been regarded to contribute to development of heart disease, weight gain and obesity.

The concentration of PUFA in milk is about 2 g/L, and the main PUFA in milk are linoleic acid (18 : 2 omega-6) and alpha-linolenic acid (18 : 3 omega-3), which may be converted to fatty acids with 20 carbon atoms, such as arachidonic acid (20 : 4 omega-6) and eicosapentaenoic acid, (EPA) (20 : 5 omega-3).

5.2.3 Minerals and Vitamins

Milk contains many minerals, vitamins, and antioxidants. The most important antioxidants in milk are the mineral selenium and vitamin E. The antioxidants have a role in prevention of oxidation of the milk, and they may also have protective effects in the milk-producing cell, and for the udder.

Milk and dairy products provide more than half of the calcium in the typical diet, and the daily intake of milk and dairy products thus has a central role in securing calcium intake. The calcium concentration in bovine milk is about 1 g/L.

Milk and dairy products are also good sources of magnesium and zinc, containing about 100 mg magnesium/L and 4 mg zinc/L. Milk also contains vitamin A, riboflavin, and vitamin B12.

6 AGRICULTURAL PRODUCTS

6.1 Introduction

Cereal food consumption has risen substantially in popularity and nutritional or functional cereal-based foods are starting to play an important role in new product development. In developing countries, the prevention of protein-energy and micronutrient deficiencies should also be of concern, but developed countries focus on deterring the alarming cases of obesity and metabolic syndrome that lead to many chronic diseases, including diabetes, high cholesterol, and hypertension.

The structure is similar in all grains. Each kernel of grain is composed of three parts, the germ, endosperm and bran. If all are presents in a grain, it is a "whole grain", such as whole wheat. When the bran or germ of the seed is removed or separated from the kernel in milling, a product is no longer "whole grain", but rather, "refined grain".

6.2 Chemical Properties and Nutritional Value

The most widely cereal-based foods used are rice, wheat and maize, which provide more than 90% of the total cereal calories, most of the protein, B-vitamins, and minerals. In some parts of the world, grains provide more than 85% of the total daily caloric intake. The inhabitants of developing countries have a higher dependency on cereal-based foods because they are cheaper, compared to animal foods.

6.2.1 Carbohydrate

Cereals are considered as an excellent source of digestible energy required for growth and work. The starch, which is almost completely digested and utilized in a normal human system, is the main calorie contributor. Another advantage is that starch releases glucose at a slower rate into the blood stream, and therefore helps in the control of diabetes.

Nowadays, whole cereals are viewed as an excellent source of dietary fiber required for the proper function of the gastrointestinal tract, and for their health-promoting effects, especially for people living in industrialized countries around the globe. The consumption of whole cereals rich in dietary fiber is even better than refined grain because of their lower glycemic index.

6.2.2 Protein

Cereals usually contain 8%~12% protein and have a good rate of protein digestion ability (80%~

90%), but unfortunately lack lysine, the most important and scarce essential amino acid in human nutrition. Therefore, protein quality of cereals is low and it is affected by the digestibility rate and mainly by the essential amino acid balance.

Nowadays, the malnutrition problems that are still widely distributed throughout the world are always observed in infants living in places where cereals provide most of the daily food intake. Aiming at the reduction of protein malnutrition, researchers are actively breeding or further improving maize, sorghum, and barley with the aim of obtaining highly sine cultivars.

6.2.3 Vitamins and Minerals

Cereals are low in fat-soluble vitamins (vitamin A, D, E and K) and devoid of vitamin C. However, they are a good source of all B-vitamins except B12. It is common to observe vitamin A deficiencies in populations that depend on cereals.

The various types of milling processes lower fiber, fat, vitamin, and mineral concentrations, and yield flours with a higher rate of starch and protein digestibility. Therefore, milling, fermentation, and cooking during the process of food production affect the composition and bioavailability of many important nutrients.

7 VEGETABLES AND FRUITS

7.1 Introduction

Because of higher fiber content and beneficial phyto-chemicals, vegetables and fruits protect against various chronic diseases, such as cardiac diseases, type II diabetes, obesity and several forms of cancer. Vegetables and fruits impart their own characteristic flavor, color, and texture to diets, and undergo changes during storage and cooking.

7.2 Chemical Properties and Nutritional Value

7.2.1 Carbohydrate

Fruits have high water content and low levels of protein and fat. The protein is concentrated in the seeds and is resistant to digestion in the small intestine and bacterial degradation in the large intestine.

However, fruits contain mostly sugars and fibers, that are extensively fermented in the large intestine, especially apples and pears, which are rich in fructose. Free fructose is poorly absorbed and would function similar to dietary fiber, escaping absorption in the small intestine while being fermented in the large intestine. Therefore, it explains why apple juice and pear juice are used to treat constipation in children.

The protein content of leaves and stems is higher than fruits and they contain low amounts of sugar. However, roots and tubers are important sources of energy as starch, which contain high amounts of sugar. On the contrary, dietary fiber of fruits is more than leaves, stems, roots, and tubers.

7.2.2 Dietary Fiber

The carbohydrate content of foods and drinks is diverse and includes digestible carbohydrates and dietary fiber. Dietary fiber is definitely an active component of fruits and vegetables and a reason for continuing to support their consumption.

Dietary fiber is listed on the nutrition fact panel and 25 g of dietary fiber is the recommended amount in a 2000-kcal diet. Therefore, whenever possible, the entire fruit or vegetable, edible peel and membrane, should be recommended to increase fiber consumption.

7.2.3 Minerals and Vitamins

Increased fruit and vegetable consumption can improve the mineral intake and reduce the risks of cardiovascular diseases and certain cancers. The main mineral elements in vegetables and fruits are potassium, magnesium, calcium, and phosphorus as well as essential trace mineral elements such as zinc, iron, and copper. In addition to mineral elements, fruits and vegetables are also recommended as a source of vitamin C and E.

Lesson 2

Vocabulary

/ Words /

diarrheal [daɪəˈri:l] adj. 腹泻的
agents [ˈeidʒənts] n. 制剂
mycotoxin [ˌmaɪkoʊˈtɒksən] n. 真菌毒素
infection [ɪnˈfekʃn] n. 传染，感染
microorganism [ˌmaɪkroʊˈɔ:rgənɪzəm] n. 微生物
fungal [ˈfʌŋg(ə)l] adj. 真菌的
calicivirus [ˈkælɪsɪˈvaɪrəs] n. 杯状病毒
Campylobacter [ˈkæmpɪləʊˌbæktə] n. 弯曲杆菌属
dysfunction [dɪsˈfʌŋkʃn] n. 机能障碍

diluent [ˈdɪljʊənt] adj. 稀释的；n. 稀释液
saturate [ˈsætʃəreɪt] vt. 使饱和；adj. 饱和的
collagen [ˈkɑ:lədʒən] n. 胶原蛋白
gelatin [ˈdʒelətɪn] n. 凝胶
nitrogen [ˈnaɪtrədʒən] n. 氮
lipid [ˈlɪpɪd] n. 脂类
fiber [ˈfaɪbər] n. 纤维
carbohydrate [ˌkɑ:boʊˈhaɪdreɪt] n. 碳水化合物
fermentation [ˌfɜ:rmenˈteɪʃn] n. 发酵

/ Phrases /

Campylobacter spp. 弯曲杆菌
hepatitis viruses 肝炎病毒
Salmonella enterica 肠道沙门氏菌
non-typhoid *Salmonella* 非伤寒沙门氏菌
heavy metals 重金属
fatty acids 脂肪酸

whole grain 全（谷）粒
biological function 生物功能
non-protein nitrogen 非蛋白氮
beta-lactoglobin β-乳球蛋白
alpha-lactalbumin α-乳白蛋白

/ Abbreviations /

saturated fatty acids (SFA) 饱和脂肪酸
phospholipid (PL) 磷脂
polyunsaturated fatty acids (PUFA) 多不饱和脂肪酸
the World Health Organization (WHO) 世界卫生组织
eicosapentaenoic acid (EPA) 二十碳五烯酸

Exercises

I. Write true or false for the following statements according to the passage

1. (　) Egg white provides all of the egg's cholesterols and contains half of the proteins.
2. (　) Shellfish provide high-quality proteins with all the dietary essential amino acids for maintenance and growth of the human body.
3. (　) The whey protein content of milk represents about 80% of milk proteins.
4. (　) Almost all of the caseins are associated with calcium and phosphate.
5. (　) Cereals are low in fat-soluble vitamins (vitamin A, D, E and K).
6. (　) Fruits have a high water content and high levels of protein and fat.
7. (　) Whey proteins are considered as rapid digested protein that give high concentrations of amino acids in postprandial plasma.
8. (　) Salt in the form of sodium chloride is commonly used in cured meats.
9. (　) Cereals are a good source of all B-vitamins and vitamin C.
10. (　) The level of nitrite or nitrate allowed in cured meats, both ingoing and residual, is strictly regulated.

II. Answer the following questions according to the passage.

1. What is the definition of foodborne disease?
2. Why the treatment on milk such as cooling and pasteurization or membrane filtration is needed?
3. What is the definition of whole grain?
4. People in developing countries have a dependency on cereals, what nutrients do they lack?
5. Why are apple juice and pear juice used to treat constipation in children?

Ⅲ. Fill in the blanks according to the passage

1. Bovine milk contains about _____ protein.
2. _____ and _____ are the major structural proteins found in the skin and bones of fish.
3. Egg white contains about half of _____ and _____ .
4. The carbohydrate content of foods and drinks is diverse and includes _____ and _____ .
5. The nitrogen in milk is distributed among _____ , _____ and _____ three categories.

Ⅳ. Translate the following words and expressions into Chinese

animal protein   food additives
environmental conditions   high-value animal protein
foodborne infections   non-protein nitrogen
heavy metals   low-fat meat products
health hazards   whey proteins

Ⅴ. Translate the following expressions into English

1. 食品科学家通常认为食品至少有六种重要属性：安全性、纯净性、使用的方便性、货架期、功能特性以及营养价值。这里我们涉及的很多重要特性在很大程度上决定了大多数消费者对食品的可接受性。

2. 人类食物供应的主要质量问题之一是其维生素和矿物质的含量，人类摄取食物从生物学角度来看是为了生存。他们在长期进化过程中已经适应了周围的环境，逐步形成了营养需求的模式。

参考译文

第2课 食品原料成分和营养

1 简介

食物中的成分可分为两大类，即活性成分和非活性成分。活性成分为身体提供能量或作为营养性食品和一些食品添加剂，非活性成分在摄入或作用于机体时不会发挥生理作用，其主要功能是充当稀释剂或促进有效成分的最终摄入或利用，如膳食纤维。本章介绍食品原料，重点阐述食品原料成分和其营养作用。

2 肉

2.1 简介

肉类是全世界范围内重要的高价值动物蛋白来源。肉和肉制品在本书被称为红肉或哺乳动物屠宰后的肌肉，包括牛肉、小牛肉、猪肉和羊肉，是饮食的一个重要组成部分。

肉和肉制品被认为是人体内胆固醇的来源，因此，在大多数发达国家肉类摄入量过高导致过高的饱和脂肪酸（SFA）的摄入。因此，营养学家和国际健康组织建议消费者可摄入瘦肉和低脂肉类制品。

2.2 肉和营养

肉由几个部分组成，肌肉纤维、结缔组织和脂肪组织等。肌肉纤维、结缔组织和脂肪在肉和鱼肉质量评定中起着关键作用。一般来说，骨骼肌包括大约90%的肌肉纤维和10%的结缔组织和其他成分。因此，富含必需氨基酸和多不饱和脂肪酸的高蛋白质肉类被认定为营养价值良好。

肉类脂肪主要包括单不饱和脂肪酸和饱和脂肪酸，主要是油酸（C18：1）、棕榈酸（C16：0）和硬脂酸（C18：0）。肉类营养价值受饱和脂肪酸（SFA）和多不饱和脂肪酸（PUFA）含量的影响。在所有物种中，大部分PUFA都存在于肌肉中的磷脂（PL）中。肉类也可能是长链$n-3$脂肪酸的重要来源，目前普遍认为其摄入量偏低。然而，肉中高水平的PUFA会导致脂质过氧化，这会影响熟肉的味道和颜色。

尽管过去20年家禽产品消费量激增，但红肉仍然占据着全球肉制品主要的市场份额。目前市场上很多是腌肉类产品，如传统的猪肉制品（火腿和培根），牛肉制品（咸牛肉）。

3 海产品

3.1 简介

对海产品需求的日益增加可能是由于其有益健康，尤其是鱼类和贝类。近年来，各种鱼类和贝类的产品如鱼皮、肌肉、鱼鳞、骨骼和内脏等都被用来提取大量的生物活性物质。鱼骨架上的鱼皮和蛋白质可作为鉴定和提取生物活性肽的廉价材料。鱼骨是钙的良好来源。

3.2 鱼和营养

胶原蛋白和明胶是鱼皮和鱼骨骼的主要结构蛋白。胶原蛋白通常用在医疗和制药工业中，尤其是作为药物的载体，这是基于其结构作用与人体的良好兼容性。明胶是一种高分子水溶性蛋白质的异质混合物，在其序列中有独特的氨基酸排列，并含有相对较高的甘氨酸、丙氨酸和脯氨酸。

鱼片加工产生的鱼架含有相当数量的肌肉蛋白质。一些鱼架作为理想蛋白来源已被用于获得功能性的水解物。

一般来说，65%~70%的鱼骨是由无机物组成的。这些无机物质是由钙、磷、氧和氢组成的羟基磷灰石。

3.3 贝类和营养

贝类提供优质蛋白质和所有膳食必需氨基酸来维持人体生理功能和生长。贝类也为健康饮食提供饱和脂肪酸、单不饱和脂肪酸和多不饱和脂肪酸。先前研究表明，从牡蛎壳和鱼胶衍生肽混合物中分离的钙化合物可作为有效的膳食钙源。

4 蛋

4.1 简介

蛋被世界卫生组织（WHO）认为是一种具有独特成分的高营养食品，是全世界的参考蛋白质。在食品生产中蛋有多种功能，因此成为许多食品配方中的首选成分。然而，多年来蛋一直受到负面宣传的影响，这与其高胆固醇含量有关。

蛋被认为是自然界最理想的食物，富含优质蛋白，不饱和脂肪酸与饱和脂肪酸的比值高，并且是矿物质和维生素的良好来源。

4.2 蛋的成分和营养价值

鸡蛋的平均质量约为57g，包括蛋黄、蛋清和蛋壳。各部分化学成分不同，如表1所示。

表1　　　　　　　　　鸡蛋的化学成分百分比

组成	总量/%	水/%	蛋白质/%	脂肪/%	灰分/%
全蛋	100	65.5	11.8	11.0	11.7
蛋清	58	88.0	11.0	0.2	0.8
蛋黄	31	48.0	17.5	32.5	2.0
蛋壳	11	—	—	—	—

资料来源：美国农业部

4.2.1 蛋黄

蛋黄占蛋总质量的30%~33%。蛋黄中含有蛋全部的胆固醇，几乎所有的脂肪和一半的蛋白质，大部分的钙、磷、铁、锌，维生素B_6、维生素B_{12}、维生素A、叶酸，一半的核黄素和硫胺素。

因此，一般来说蛋黄的营养密度比蛋清要高，含有几乎所有维生素（除了维生素C）。蛋黄中含有甘油三酯、磷脂和固醇。主要的磷脂是磷脂酰胆碱（卵磷脂）。蛋黄中的蛋白质主要是卵黄蛋白，存在于脂蛋白和脂卵黄蛋白中。

4.2.2 蛋清

蛋清含有大约一半的蛋白质和核黄素。蛋清占鸡蛋总质量的55%~60%。

4.2.3 气室、膜和外壳

气室是由两个壳膜分离而形成的，通常在鸡蛋的大头末端，在冷却过程中鸡蛋内容物收缩形成。蛋壳的下一层是内层和外膜。这些相对较薄的角蛋白样膜是抵御细菌入侵的主要防御屏障。

蛋壳占鸡蛋总质量的 9%~12%，主要是碳酸钙（94%）和一些碳酸镁（1%）、磷酸钙（1%）和有机物，有机物主要是蛋白质（4%）。

5 乳和乳制品

5.1 简介

乳和乳制品含有大量的必需营养素，有悠久的食用历史。人类仅食用个别少数动物的乳，如奶牛、水牛、山羊和绵羊，其中主要是奶牛。

乳中含有水、脂类、碳水化合物、蛋白质、盐和一些其他成分。乳中含有青少年生长发育所必需的营养物质，是脂质、蛋白质、氨基酸、维生素和矿物质的良好来源，同时也含有免疫球蛋白、激素、生长因子、细胞因子、核苷酸、多肽、多胺、酶和其他生物活性肽。

5.2 乳成分和营养价值

5.2.1 蛋白质

牛乳含有蛋白质约 3.3 g/100g（表 2）。牛乳蛋白具有较高的生物学价值，因此牛乳是必需氨基酸的良好来源。此外，牛乳含有大量具有生物活性的蛋白质，既作为抗菌物质也可促进营养素的吸收，以及作为生长因子、激素、酶和抗体。

表 2　　　　　　　　　　　山羊乳、牛乳和人乳的组成

成分	山羊乳	牛乳	人乳
脂肪/（g/100g）	3.8	3.6	4.0
蛋白质/（g/100g）	3.5	3.3	1.2
乳糖/（g/100g）	4.1	4.6	6.9
固形物/（g/100g）	12.2	12.3	12.3
能量/（cal/100g）	70	69	68

乳中的氮分布在酪蛋白、乳清蛋白和非蛋白氮中。乳蛋白的含量是通过分析氮元素（N）含量而得出。氮是乳中主要成分，占蛋白质量的 1/6 左右。由于各类乳蛋白中氮元素的含量几乎一致，为 16%，因此，常使用氮含量乘 6.25 得出乳总蛋白质的含量。

牛乳中的酪蛋白含量占牛乳蛋白的 80%，几乎所有的酪蛋白都与钙和磷酸盐有关。因此，酪蛋白的生物学功能是提供钙和磷酸盐。

另一种蛋白叫作乳清蛋白，包括多种蛋白。乳清蛋白是牛乳生产乳酪后排出的液体，添加于多种食品中。乳清蛋白是球状蛋白质，比酪蛋白更易溶于水，其主要为 β-乳球蛋白、α-乳

清蛋白、牛血清白蛋白和免疫球蛋白。乳清蛋白是易于消化的蛋白质,食用后血浆氨基酸的浓度提高。

5.2.2 脂质

牛乳含有约 3.6g/100g 的脂质,山羊乳为 3.8 g/100g。乳中的脂肪球以稳定的乳浊液形式存在,脂肪球均匀分散在含非脂乳固体的液相中。越小的脂肪球越有利于消化。山羊乳、牛乳、水牛乳和绵羊乳脂肪球的平均直径分别为 3.5μm、4.5μm、5.9μm 和 3.3μm。

甘油三酯约占脂类的 95%,由不同长度的脂肪酸(4~24 个 C 原子)和饱和脂肪酸组成。因此,富含乳脂的饮食有助于提高总膳食脂肪酸的比例。

超过一半的牛乳脂肪酸是饱和脂肪酸。饱和脂肪酸月桂酸、肉豆蔻酸(14∶0)和棕榈酸(16∶0)会提高血液胆固醇水平。富含饱和脂肪酸的饮食会导致心脏病、体重增加和肥胖。

牛乳中 PUFA 的浓度约为 2g/L,牛乳中主要的 PUFA 是亚油酸(18∶2,ω-6)和亚麻酸(18∶3,ω-3),可转化为 20 个碳原子的脂肪酸,如花生四烯酸(20∶4,ω-6)和二十碳五烯酸(EPA)(20∶5,ω-3)。

5.2.3 矿物质和维生素

牛乳含有多种矿物质、维生素和抗氧化剂。牛乳中最重要的抗氧化剂是矿物元素硒和维生素 E。抗氧化剂起到防止牛乳氧化的作用,也在奶牛乳房细胞和乳房中起保护作用。

乳和乳制品提供超过一半的膳食钙,每天饮用乳和乳制品可保证钙的摄入。牛乳中钙的浓度约为 1g/L。

乳和乳制品也是镁和锌的良好来源,含有大约 100 mg/L 的镁和 4 mg/L 的锌。牛乳也含有维生素 A、核黄素和维生素 B_{12}。

6 农产品

6.1 简介

谷物食品消费量大幅上升,营养性或功能性谷物食品开始在新产品开发中发挥重要作用。在发展中国家,预防蛋白质-能量和微量营养素缺乏的问题依然值得关注。但是,发达国家应该重点关注导致多种慢性疾病(糖尿病、高血脂和高血压)的肥胖和代谢综合征病例。

所有谷物的结构都是相似的。每粒谷物都由三部分组成:胚芽、胚乳和麸皮,含有这三部分的谷物,就是"全谷物",如全麦。当碾磨加工时,麸皮或胚芽与谷物分离或去除,此时不再是"全谷物",而是"精加工谷物"。

6.2 化学性质和营养价值

最普遍食用的谷物是大米、小麦和玉米,提供了超过 90% 的谷类热量、大部分蛋白质、B 族维生素和矿物质。在世界一些地方,谷物提供了超过 85% 的每日热量摄入。发展中国家的居民摄入更多的谷物类食品,因为与动物性食品相比,谷物类食品更便宜。

6.2.1 碳水化合物

谷物被认为是生长、工作所需消化能的极好来源。在正常的人体消化系统中,淀粉几乎完全被消化和利用,是主要的热量来源。另一个好处是,淀粉会以较慢的速度释放葡萄糖到血液中,从而利于调控糖尿病。

目前,全谷物食品被视为一种极好的膳食纤维来源,用于调控胃肠道的正常机能及促进人体健康,尤其是对全球工业化国家的人们而言。由于血糖指数较低,食用富含膳食纤维的全谷物食品比精加工谷物要好。

6.2.2 蛋白质

谷物通常含有8%~12%的蛋白质,并且具有良好的蛋白质消化率(80%~90%),但是缺乏赖氨酸,赖氨酸是人类营养中最重要且最易缺乏的必需氨基酸。因此,谷物的蛋白质品质较低,受蛋白质的消化率和必需氨基酸平衡的影响。

如今,世界各地仍广泛存在营养不良问题,尤其是以谷物为主要膳食的婴幼儿营养不良更为突出。为了减少蛋白质营养不良,研究人员正在积极培育,进一步改良玉米、高粱和大麦,以期获得高质量的品种。

6.2.3 维生素和矿物质

谷物中脂溶性维生素(维生素A、维生素D、维生素E和维生素K)含量低,也缺乏维生素C。但是除了维生素B_{12},谷物是其他B族维生素的良好来源。以谷物为主要膳食结构的人群极易缺乏维生素A。

各种碾磨加工处理降低了谷物纤维、脂肪、维生素和矿物质的含量,提高了淀粉和蛋白质的消化率。因此,在食品加工过程中,碾磨、发酵和烹饪影响了重要营养成分的组成和生物利用率。

7 蔬菜和水果

7.1 简介

由于富含高纤维和有益的植物化学物质,蔬菜和水果可预防多种慢性疾病,如心脏病、Ⅱ型糖尿病、肥胖症和多种癌症。蔬菜和水果将其风味、色泽和质地赋予食物中,在储藏和烹饪加工过程中发生变化。

7.2 化学性质和营养价值

7.2.1 碳水化合物

水果含水量高,蛋白质和脂肪含量低。水果蛋白质集中在种子中,可耐受小肠的消化和大肠的细菌降解作用。

然而,水果含有大量的糖和纤维,可在大肠中发酵,尤其是苹果和梨,富含果糖。游离果糖不易被吸收,其功能类似于膳食纤维,在大肠内发酵时,可以逃避小肠的吸收。因此,这就是苹果汁和梨汁可用来治疗儿童便秘的原因。

叶和茎的蛋白质含量高于水果，但含糖量低。然而，根和块茎是淀粉的重要来源，含有大量的糖。相反，水果的膳食纤维比叶、茎、根和块茎都要高。

7.2.2 膳食纤维

食物和饮料的碳水化合物种类繁多，包括可消化的碳水化合物和膳食纤维。膳食纤维是水果和蔬菜的有效成分，是人们食用蔬菜水果的主要原因。

膳食纤维应在营养成分表上列出，推荐摄入量2000kcal的饮食中应含25 g的膳食纤维。因此，完整的水果、蔬菜或可食的果皮、外膜均推荐食用，以增加膳食纤维的摄入。

7.2.3 矿物质和维生素

增加水果和蔬菜的摄入可提高矿物元素的摄入，减少心血管疾病和某些癌症风险。蔬菜和水果中的矿物元素主要是钾、镁、钙和磷，以及锌、铁和铜等必需微量元素。除了矿物质元素外，水果和蔬菜也是维生素C和维生素E的良好来源。

LESSON 3

ISSUES IN FOOD SAFETY

1 INTRODUCTION

In recent years, with the rapid development of economy, people's living standards are continuously improving, and more and more people are paying attention to food safety issues, which also have become a global public health issue. Otherwise, the global food supply chain has changed, which means that our food on the table can possibly come from any corner in the world. Therefore, there are no boundaries for food safety issues; people from all over the world are trying their best to solve it. Although the governments of the entire world have been taking a series of effective measures to reduce and prevent their occurrences, food safety accidents happened frequently each year around the world. For example, in 1996, the outbreak of mad cow disease in Britain; in 1997, bird-flu in Hong Kong; in 1998, swine encephalitis in Southeast Asia; in 1999, dioxin in Belgium and other countries; in 2001, the outbreak of foot-and-mouth disease in Europe; in 2008, the shocking "Sanlu milk powder incident" in China; in 2009, global influenza A (H1N1) influenza; in 2009, the "lean meat" event in China; and in 2011, ditch oil accident in China. These food safety issues not only have endangered the physical and mental health of consumers, but also have brought incredible loss to national economy. According to the national food poisoning situation published by the National Ministry of Health, only in 2003, there were 379 major food poisoning incidents in China, resulting in 12,876 people poisoned, and 323 deaths. In 2007, there were 506 major food poisoning incidents nationwide, with 13,280 people poisoned and 258 deaths. In 2008, 437 incidents of heavy food poisoning occurred throughout the country and led to 13,325 people poisoned and 152 deaths.

Food safety issues also reflected that there were many hidden dangers in today's food products, as the main reasons include: inferior raw materials used in the food process, the addition of toxic substances, overuse of food additives, abuse of non-food processing chemical additives. The safety situations of agricultural products and poultry products are also not optimistic. Antibiotics, hormones and other harmful substances remain in poultry, livestock and aquatic products. Genetically modified foods also have potential threats to humans, though there is not enough evidence to prove that genetically modified foods are harmful to humans. Of course, food safety problems are also caused by some social reasons, for example some enterprises ignore food safety and product quality in order to gain more profits;

enterprise management confusion; new food quality and safety problems caused by emerging technologies; imperfection of social supervision systems and laws and so on.

At present, how to provide safe food for consumers is still a major challenge for some developed and developing countries, which is also the focus of policy-makers all over the world. To solve food safety problems, it is necessary to know the possible links of food safety issues' occurrence and the factors that affect food safety. At last, the occurrence of food safety problems can be reduced or even eliminated by enacting laws and strengthening supervision. This chapter mainly analyzes the possible food safety problems that occur in the whole food supply chain. It introduces important food safety accidents that have occurred in recent years all over the world. The objectives are to make people understand the possible links of food safety issues' occurrence from planting to table food and reduce the occurrence of food safety issues as much as possible. Finally, some corresponding measures are put forward as a response to the food safety issues of the food supply chain.

2 FOOD SAFETY ISSUES ORIGINATING FROM THE ENTIRE FOOD SUPPLY CHAIN

Food safety issues originate from the entire food supply chain, and all the links, including materials, growth environment, climate and soil, processing, packaging, storage, transportation and sale, are possible sources of food safety issues. Figure 1 summarizes the possible food safety issues and risks in the food supply chain. These issues and risks are largely due to human activities, which present a high degree of complexity and uncertainty, and are also directly related to the types of food and food attribute and the types of substance, technology, and region, etc.

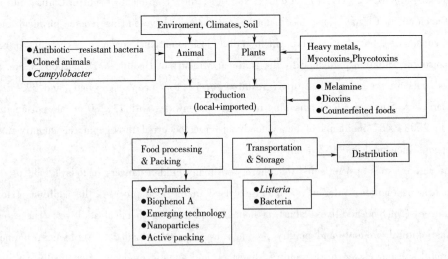

Figure 1 Possible food safety issues and risks in food supply chain

2.1 Environment, Climate and Soil

The sources of food safety issues come from the growth environment of food process materials, climate and soil, which are closely affected by natural disasters, pollution and climate. Figure 2 clearly shows the effects of environment, climate and soil on food contamination.

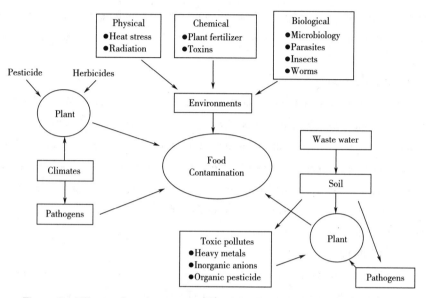

Figure 2 Effects of environment, climate and soil on food contamination

2.1.1 Environment

Pollution from human activities on environment and biosphere is posing a steadily rising threat to maintenance of life in the world. Various environmental pollution resulted in food contamination, which plays an ultimate role in damaging the human genetic material DNA. The pollution of human beings to the environment and the threat of the environment to human beings occur simultaneously. Household waste, animal feed, and plant fertilizers are serious carriers of environmental pollutants and food contaminants. The effects of these factors on environment including physical factors such as heat stress (heat waves and cold waves) and irradiation; chemical factors such as plant fertilizers and toxic substances; biological factors such as microorganisms, parasites, insects and worms, etc. Environmental pollution also affects conversely the atmosphere, soil, water, animals, plants and microbes, which will eventually result in food safety issues.

2.1.2 Climate

The potential impact of climate change on food safety is a widely debated and investigated issue. Climate change impacts mycotoxins' formation in plant products in the field or during storage. Climate change also causes a large number of insects to reproduce, thus affecting the growth of plant crops and the growth of weeds in the field. Climate change leads to a large number of pesticides and insecticides being used, and the uses of pesticides and herbicides cause pesticide residues as well as harmful chemical residues in crops. The pathogenic bacteria reproduce quickly under extreme weather conditions, such as flooding and heat waves. However, the consequences of climate change are the transformation of the food system, which comprises all the stages from "farm to fork" (mainly primary production, processing, transport and trading).

2.1.3 Soil

The crops can hardly grow without soil, the quality of which is directly related to the safety of food raw materials. At present, domestic wastewater and untreated industrial wastewater are the main sources of soil pollution.

2.1.3.1 Pathogen Contaminants

Domestic wastewater, sewage from livestock and poultry farm, leather industry, slaughterhouse and wastewater from hospitals often contain various pathogens, such as viruses, pathogens and parasites. These pathogens are retained in soil and transferred from soil to crops, causing food safety problems. The soil contaminated by the pathogens can lead to the rapid growth of microbes, many of which are pathogenic bacteria, insects, worms and viruses. They often coexist with other bacteria and *Escherichia coli*. Pathogen contaminants can transfer from soil to the crops by various pathways. Once the conditions are suitable, they will cause human illness.

2.1.3.2 Toxic Pollutants

Toxic pollutants refer to the accumulation of living organisms to a certain quantity that makes body fluid and tissues undergo biological and physiological changes in their functions, causing a temporary, persistent, or even life-threatening pathological state, such as heavy metals and decomposable organic pollutants. The toxicity of pollutants is closely related to the amount of intake. Soil toxic pollutants caused by the discharge of domestic sewage and industrial wastewater mainly include: ① Heavy metals. It has been long recognized that heavy metals are among the major contaminants of food supplies and may be considered as the most important problem damaging our environment. Crops can absorb heavy metals from contaminated soil and atmosphere, which can be accumulated for a long time. A large amount of heavy metals can lead to serious food safety problems. Heavy metals such as Cd and Pb have been shown to have carcinogenic effects, and high concentrations of Cu, Cd, and Pb in fruits and vegetables are related to the high prevalence of upper gastrointerestinal cancer. Heavy metals do not easily disappear in nature; they can be accumulated through the food chain. Heavy metals accumulate in human tissues. For example, arsenic, chromium, cadmium, and lead affect the nerve system and kidneys, as well as the hematopoietic system (blood). Cadmium and chromium affect liver, and they harm men's reproductive system. In addition to directly causing diseases in humans, some heavy metals may also promote the development of chronic diseases. ② Inorganic anion includes carcinogens such as NO、F and CN anions. Toxic cyanides are mainly from industrial wastewater discharges. These inorganic anions directly and indirectly affect crops and result in food safety issues. ③ Organic pesticides and polychlorobiphenyls. At present, there are almost 6000 kinds of organic pesticides in the world, and more than 200 are common pesticides. Pesticides sprayed on farmland enter the soil through leaching, resulting in pollution. Organic pesticides can be divided into organophosphorus pesticides and organochlorine pesticides. Although organophosphorus pesticides have very high toxicity, they are generally easy to degrade and difficult to accumulate, so their effects on the ecosystem are not obvious. The majority of organochlorine pesticides are highly toxic, difficult to degrade, highly accumulative and have a significant impact on the ecosystem. Crops can also absorb organic pesticides from contaminated soil and cause pesticide residues, which can lead to serious food safety issues by a long-term accumulation.

2.2 Animals and Plants

Animals and plants are food process materials, which grow under the conditions of environment, climate and soil. These conditions are related to the quality of food process materials and also related to food safety issues. For plants, some chemical toxic substances such as heavy metals, inorganic anions and organic pesticides, result in food safety issues. Some biological substances such as mycotoxins and phycotoxins also cause the occurrence of food safety issues. The factors affecting animal safety include the use of antibiotics, cloned animals, and pathogenic *Campylobacter* parasites in the intestines of mammals and birds.

2.3 Food Processing and Packaging

The problems of food materials in food processing, such as melamine addition, dioxin, and food processing from inferior raw material, lead to frequent food safety issues. Food safety issues that occur in the processing of food include incomplete physical removal of contaminants, incomplete inactivation of micro-organisms, spores, viruses, parasites (special risks for the chilled storage of foods of animal origin), incomplete inactivation of natural anti-nutrients, allergens, and toxins. Nutrient loss or decrease in nutrient bioavailability is because of over-processing, as well as formation of toxic molecules (acrylamide, heterocyclic amines, HMF) and enhancement of microbial growth (slicing, grinding, long storage). Toxic chemical substances from packaging materials are transferred from package materials to food. The following are the main substances that may occur in food safety issues during the process of food processing and packaging.

2.3.1 Toxic Substances in Food Processing and Packaging
2.3.1.1 Acrylamide

Acrylamide (molecular formula: C_3H_5NO) can be formed in some carbohydrate-rich foods, such as potatoes and cereal-based products (potato chips, French fries, bread, crisp-breads, biscuits, breakfast cereals, coffee), and these products can possibly form acrylamide during cooking processes such as frying, baking, roasting, or microwaving over 120℃. Acrylamide levels are observed up to 5mg/kg in potato chips and 2mg/kg in biscuits. Acrylamide may also be formed in heated fat-rich dry foods such as nuts. At present, the exact mechanisms of its formation remain unsure, but it involves the Maillard reaction between asparagine and reducing sugars. Acrylamide is also a highly reactive industrial chemical substance used for the production of plastics and materials. It may migrate into foods from some food contact materials. Acrylamide is a known carcinogen to experimental animals, and may be a reproductive toxicant. The link between acrylamide and the risk of cancer in humans is uncertain, although people have been exposed to acrylamide in their diets for a long time. Recent studies suggest that tolerable daily intake of acrylamide for carcinogenic levels should be set at 2.6mg/kg b.w. per day, and Canadian, Swedish and US studies estimate the average exposure of adults to acryl amide in foods is 0.3, 0.5 and 0.4mg/kg b.w. per day, respectively.

2.3.1.2 Bisphenol A

Bisphenol A (BPA) (4,4-dihydroxy-2,2-diphenylpropane) is used, usually in combination with other chemicals, to manufacture plastics and resins. BPA is thus present in polycarbonate food containers such as returnable beverage bottles, infant feeding bottles, tableware, and storage containers. BPA residues are also present in epoxy resins used to make protective coatings for food and beverage cans and vats. During food storage, BPA may migrate in small amounts into food and drinks. Bisphenol A has side effects on female genitals. Concerning the safety of BPA, European Food Safety Authority (EFSA) sets a tolerable daily intake (TDI) of 50 μg/kg b.w. per day. EFSA notes that newborn infants', children's, and adults' intake of BPA is well below the TDI. There are strong controversies about various possible toxic effects of BPA. BPA is permitted in food contact materials in EU, USA, Japan, but some US states or EU countries have already banned its use for infant food containers.

2.3.2 Fraud and Adulterated Foods

Profit-based criminal frauds have caused a number of food crises through direct or indirect addition of toxic substances to foods or feeds. The following are some food safety issues about fraud and adulterated food.

2.3.2.1 Fraud and Adulterated Oils

Ukrainian sunflower oil containing mineral oil, was exported to Europe, which led to withdrawal from EU markets. Spanish oil syndrome, with 20,000 persons affected, and over 800 deaths, was due to a denatured rapeseed oil, which was sold as edible olive oil. In 2011, the ditch oil accident happened in China. Ditch oil generally refers to all kinds of inferior oil that exist in life, such as recovered edible oil, and repeated use of fried oil. The largest source of ditch oil is from hotel and restaurant drainage pipes for human consumption, and long term consumption may cause cancer and great harm to the human body. Its main components are still triglycerides, but it has much more pathogenic and carcinogenic toxins than real edible oils.

2.3.2.2 Melamine

In 2008, melamine milk case happened in China, and about 300,000 infants and young children drank infant milk containing melamine, which caused kidney and urinary tract diseases, including kidney stones, with six reported deaths. Investigations showed that melamine had been deliberately added to dilute raw milk to boost its apparent protein content. Because melamine is cheap and easily accessible, there is an economic incentive for its illegal addition. The molecular formula of melamine is $C_3H_6N_6$. Nitrogen level is very high in melamines, accounting for 66% of the total mass. But melamine is present in nonproteinaceous molecules. Thus, the addition of melamine can lead to an incorrectly high protein reading. The acuteness of melamine's toxicity was low, and the median lethal dose was 3,000mg/kg body weight in mice and rats. When the dose increased up to 18,000mg/kg feed, the toxicity was found to affect the excretory organs, kidney and bladder, and these effects were more obvious in male organs than in female.

2.3.2.3 Dioxins

Dioxins are colorless, tasteless, fat soluble substances with serious toxicity. In 2008, Rapid Alert System for Food & Feed (RASFF) sent alert notifications for dioxins (present above legal limits) in foods. An example is that of pork from Ireland, with a large trace and recall operation (from 54 countries) of pig meat and many processed pork products (RASFF annual report 2008). Dioxins were found in pig meat at the level that is 100 times as much as the EU maximum limit of 1pg/g fat. The sources of this contamination/adulteration was investigated as follows. Pigs from about 50 farms were fed with contaminated bakery wastes. These wastes consisted of bread crumbs which had been dried by direct contact with combustion gases. And the fuel burned to make the combustion gases had been contaminated by adding oil illegally.

2.3.3 Food Additives

The main categories of additives are colors, preservatives, antioxidants, emulsifiers and stabilizers, and sweeteners. Some food additives are neurotoxic, which means they are capable of altering the normal activity of the nervous system, and even killing neurons. The illegal production of foods with the use of non-authorised food or feed additives is not a single case. In 2010, some Chinese wheat flour was adulterated with pulverised lime added with bleaching agent.

2.3.4 Fraud, Counterfeit Foods and Drinks

According to the statistics in 2009, the global level of fraud as counterfeiting in the food and drink industry is estimated at about $50 bn/year. The most frequently counterfeit foods and drinks are: fruits, conserved vegetables, baby food, milk powder, butter, instant coffee, spirits, drinks and confectionaries. Many counterfeit foods and drinks obtain high profits through damage to famous brands. Food safety issues are caused by the quality of counterfeit products.

2.3.5 Emerging Food Processing Technologies

Some emerging food processing technologies will also pose a potential threat to food safety, and some technologies are also widely used in the food industry. The following is a brief introduction to some new technologies with possible causes of food safety issues.

2.3.5.1 New Physical Treatments

High pressure pasteurization: incomplete microbial inactivation and spore inactivation, some chemical reactions.

Pulsed electric fields: incomplete spore inactivation, electrochemical reactions; metal transfer from electrodes.

Cold plasma: free radicals, oxidations.

Light pulses and UV: photo-oxidative reactions.

Ultrasound: incomplete microbial and spore inactivation.

Ohmic heating: metal transfer from electrodes.

Ionising radiation: free radicals, oxidative reactions.

2.3.5.2 New Chemical Treatments

Modified atmosphere: it can delay microbial growth, but tends to over-extend shelf-life and storage time.

Anti-microbial agents: incomplete microbial and spore inactivation; microbial growth only delayed in storage, potential toxicity.

2.3.6 Nanotechnologies

Engineered nanomaterials (ENM) or nanoparticles are already used in ingredients, additives, fertilizers, pesticides, drugs and packaging materials. ENM have specific physico-chemical properties. The following are the main uses of nanotechnologies in food.

(1) as food ingredients and additives: increase solubility and dispersibility, stabilize emulsions without emulsifier, improve texture.

(2) as delivery systems for bioactive compounds: protect targeted delivery of nutrients; improve the availability of bioactive compounds.

(3) in innovative packaging: anti-bacterial films.

The safety issues caused by the application of nanotechnology in food industry are still being discussed and studied, such as the detection, properties and dosages of nanomaterials, their absorption, metabolism, and excretion, their toxicology and environmental impact. All of these are waiting on safety assessment and authorization.

2.3.7 Some Additional Challenges

Food safety issues are facing the contradictions between the need of innovative technologies and food safety issues. In spite of EFSA and food law, stricter safety and quality standards, intensified quality control and monitoring, the number of reported food safety incidents has increased, and consumers' trust on food safety has decreased. Some agro-food technologies tend to elicit consumer rejection:

Ionising irradiation of foods;

Hormonal (and antibiotic) treatment of animals to hasten growth and increase meat or milk production (banned in the EU);

Various food additives;

Excessive use of crop fertilizers and pesticides;

Genetically modified food (GMF) crops and food ingredients;

Genetically modified animals (including cloned animals).

There are indeed challenges in matching the fast pace of innovation in food production and processing with risk assessment methodologies (pathogen testing, allergen testing, toxicology evaluations, and environmental impact).

The consumer's "right to informed choice" is well established in the EU. Mandatory labeling for irradiated foods and GM foods has discouraged manufacturers and retailers to place such foods on the market, despite their potential advantages.

2.3.8 Active Food Packaging

Active food packaging refers to what can supply the information of packaged food properties during the circulation and storage periods by detecting the environment conditions of packaged food. Such as time-temperature indicator, ripeness indicator, and packaging leakage, etc. What active food package brings safety risk to food safety issues is that migration of chemicals and their degradation products into the food. Nano particles could increase risks. European Food Safety Authority (EFSA) has an authorization scheme, focusing on migration data of chemicals and their toxicological properties .

2.4 Food Transportation and Storage

In the processes of food transportation and storage, if the management of refrigerated distribution channels is improper, food will be contaminated by some microorganisms. For some ready-to-eat foods, the growth of *Listeria monocytogenes* is very fast if pH > 4.5 and/or Aw > 0.91. Some major outbreaks are associated with fresh vegetables and fruits, such as salad greens, lettuce, sprouts, and melons in which most frequent pathogens are: *Norovirus*, *Salmonella* and *E. coli* O157.

2.5 Distribution and Sales of Food

The distribution and sales of food are the same as the transportation and storage of foodstuffs, and will lead to microbial contamination if they are not handled properly.

3 COUNTERMEASURES AGAINST FOOD SAFETY ISSUES

Food safety issues may occur in any link of the food supply chain. If any link is monitored and managed improperly, food safety issues will occur. Therefore, controlling and reducing the occurrence of food safety issues is a great challenge for countries all over the world. To solve the possible problems of food safety in the food supply chain, some corresponding countermeasures have been put forward.

3.1 Establishment of Reasonable Food Safety Management Tools

Reasonable food safety management tools play a critical role in the prevention of food safety problems. These tools include food legislation, national and international standards, quality management systems, risk analysis, risk-based inspections and controls, monitoring and alert systems for food contaminants and quantitative microbial risk assessment, nutrition and toxicology studies, and elaborate food processing technologies. These tools have brought consumers the right to make informed choices about safe foods.

3.2 Establishment of Predictive, Early Warning and Communication Systems

Establishment of predictive, early warning and communication systems can make preventive measures possible before food hazards become real risks.

3.3 Supervision of Every Link of Food Supply Chain

Biological and chemical pollution may occur in the processing, transportation, storage, and distribution of food raw materials and need to be avoided. As evidenced by recent events or crises, numerous pathogen outbreaks, such as dioxin, acrylamide, melamine, and bisphenol A cases, are still the main reasons for many food safety issues. To reduce and eliminate these issues, every link in the food supply chain needs to be strictly monitored.

3.4 Establishment of Strict Market Regulation System

The melamine case, the international and national counterfeited foods and drinks are all profit-motivated frauds. These adulteration incidents pose main and steady threats to food safety, which need strict market regulation and supervision.

3.5 Establishment of Safety Assessment and Pre-market Authorization Procedures

The use of innovative technologies in food processing are still being debated because of its impact on food safety. Safety assessment and authorization procedures need to be applied to the innovative technologies to relieve consumers' worries and allow them to choose safe food.

⫅))) Lesson 3

Vocabulary

/ Words /

antibiotic [ˌæntibaɪˈɑːtɪk] n. 抗生素
counterfeit [ˈkaʊntərfɪt] n. 伪造品
emerging [iˈmɜːrdʒɪŋ] adj. 新兴的
simultaneously [saɪməlˈteɪniəsli] adj. 同时的
contamination [kənˌtæmɪˈneɪʃn] n. 污染
fertilizer [ˈfɜːrtəlaɪzər] n. 肥料
accumulation [əˌkjuːmjəˈleɪʃn] n. 积累
allergen [ˈælədʒən] n. 过敏原
outbreak [ˈaʊtbreɪk] n. 爆发
authorization [ˌɔːθərəˈzeɪʃn] n. 授权

melamine [ˈmeləmiːn] n. 三聚氰胺
dioxin [daɪˈɑːksɪn] n. 二噁英
acrylamide [əˈkrɪləmaɪd] n. 丙烯酰胺
pathogen [ˈpæθədʒən] n. 病原菌
pesticide [ˈpestɪˌsaɪd] n. 农药
fraud [frɔːd] n. 伪劣品
parasite [ˈpærəsaɪt] n. 寄生虫
virus [ˈvaɪrəs] n. 病毒
gastrointestinal [ˌɡæstroʊɪnˈtestɪnl] adj. 胃与肠的

/ Phrases /

food supply chain 食品供应链
effective measures 有效措施
environmental pollution 环境污染
human genetic material 人类遗传物质

domestic wastewater 生活污水
heavy metal 重金属
organic pesticide 有机农药
pesticide residue 农药残留
organophosphorus pesticide 有机磷农药
organochlorine pesticide 有机氯农药
a long-term accumulation 长期积累
food process material 食品加工原料
anti-nutrients 抗营养因子
incomplete inactivation 不完全灭活
adulterated oil 掺假油
ditch oil 地沟油
emerging food technology 新兴食品加工技术
electrochemical reaction 电化学反应
photo-oxidative reaction 光氧化反应
ohmic heating 欧姆加热
anti-microbial agents 抗菌剂
anti-bacterial film 抗菌膜
engineered nanomaterial 工程纳米材料
farm to fork 农田到餐桌
right to be informed choice 知情权

/ Abbreviations /

European Food Safety Authority (EFSA) 欧洲食品安全局
tolerable daily intake (TDI) 日允许摄入量
Rapid Alert System for Food & Feed (RASFF) 食品和饲料快速预警系统
engineered nanomaterials (ENM) 工程纳米材料
genetically modified food (GMF) 转基因食品

Exercises

I. Write true or false for the following statements according to the passage.

1. ( ) The global food supply chain has changed, which means that our food on the table can possibly come from any corner in the world.

2. ( ) Food safety issues occurred in the entire food supply chain.

3. ( ) Climate change impacts mycotoxins form in plant products in the field or during storage. Climate change also causes a large number of insects to propagate, thus affecting the growth of plant crops.

4. ( ) The consequences of climate change in the food system, which comprises all the stages from "farm to fork".

5. ( ) At present, domestic wastewater and untreated industrial wastewater are the main causes of soil pollution.

6. ( ) Toxic pollutants refer to the accumulation in living organisms to a certain quality can make body fluid and tissues take biological and physiological functional changes, causing a temporary or persistent pathological state, and even threaten life, such as heavy metals and decomposable organic pollutants.

7. ( ) It has been long recognized that heavy metals are among the major contaminants of food supply and may be considered as the most important problem to our environment.

8. (　) Arsenic, chromium, cadmium, leads affect the nervous system and kidneys, does not harm hematopoietic system (blood).

9. (　) Crops can also absorb organic pesticides from contaminated soil and cause pesticide residues, which can lead to serious food safety issues by a long-term accumulation.

10. (　) Acrylamide is also a highly reactive industrial chemical used for the production of plastics and materials. It may migrate into foods from some food contact materials. Acrylamide is a known carcinogen to experimental animals, and may be a reproductive toxicant.

11. (　) The link between acrylamide and cancer risk in humans is certain.

12. (　) European Food Safety Authority (EFSA) set a tolerable daily intake (TDI) that intakes of newborn infants, children and adults were of 50μg/kg b. w.

13. (　) The most frequently counterfeited foods and drinks are: fruits, conserved vegetables, baby food, milk powder, butter, instant coffee, spirits, drinks, confectionaries.

14. (　) Some emerging food processing technologies will not pose a potential threat to food safety issues, and some technologies are also widely used in the food industry.

15. (　) The number of reported food safety incidents has increased, and consumers trust in food safety has decreased.

16. (　) Melamine is present in nonproteinaceous molecule. Thus, the addition of melamine can lead to an incorrectly high protein reading.

17. (　) Because of potential advantages, mandatory labeling for irradiated foods and GM foods has encouraged manufacturers and retailers to place such foods on the market.

18. (　) Active food package bring safety risk to food safety issues is that migration of chemicals and their degradation products into the food.

19. (　) Some ready-to-eat foods, the growth of *Listeria monocytogenes* is very fast if pH > 4.5 and/or Aw > 0.91.

20. (　) Salad greens, lettuce, sprouts, and melons are leading vehicles, with *Norovirus*, *Salmonella* and *E. coli* O157 as most frequent pathogens.

Ⅱ. Answer the following questions according to the passage.

1. Food safety issues reflected many hidden dangers in today's food safety, what reasons are including?

2. Food safety issues occurred in the entire food supply chain, what links possible resulted in food safety issues?

3. Which threats will pose by emerging food processing technologies to food safety issues?

4. Which agro-food technologies are tending to elicit consumer rejection?

5. What does active food package refer to? Which safety risks will active food package bring to food safety issues?

Ⅲ. Fill in the blanks according to the passage.

1. There are many factors affected on environment pollutions including physical factors, such as _____ and _____; chemical factors, such as _____ and _____; biological factors, such as _____, _____, _____ and _____.

2. Soil toxic pollutants caused by the discharge of domestic sewage and industrial wastewater mainly include _____, _____ and _____.

3. Some toxic substances present in food process and package may cause food safety risks, mainly included _____ and _____.

4. Acrylamide may be formed in carbohydrate-rich foods, _____ and _____ during cooking processes such as frying, baking, roasting, microwaving over 120℃.

5. Engineered nanomaterials (ENM) or nanoparticles are already used in ingredients, additives, _____, _____ and _____.

Ⅳ. **Translate the following words and expressions into Chinese.**

food safety accidents   genetically modified food
non-food processing chemical additives   right to be informed choice
food supply chain   active food package
domestic wastewater   ready-to-eat foods
emerging food technologies   pre-market authorization procedures

Ⅴ. **Translate the following expressions into English.**

1. 他们还就诸如食品添加剂、化学品和微生物污染物以及农用化学品残留物等众多食品安全问题提供科学建议。

2. 食品安全问题与人们的日常生活息息相关，稍不留心便会酿成无法弥补的大祸。

3. 随着经济的发展，食品消费日趋多元化，以及消费者对食品安全问题越来越关注，这对食品物流业提出了较高的要求。

4. 在欧盟，消费者对产品的知情选择权已经建立，辐照食品和转基因食品的强制性标签阻碍了制造商和零售商将这些产品推向市场。

参考译文

第3课 食品安全问题

1 简介

近些年来，随着经济的高速发展，人们的生活水平不断提高，食品安全问题日渐成为人们关注的焦点，并发展成为一个全球性的公共卫生问题。另外，全球食品供应链正发生着改变，这意味着我们餐桌上的食物可能来源于世界的任何一个角落。因此，食品安全问题无国界，是全世界人们亟待共同解决的问题。虽然世界各地的政府部门都在采取一系列的有效措施来减少和预防食品安全事故的发生，但是在世界各地，每年的食品安全事故还是频繁发生。例如，1996年英国爆发的疯牛病，1997年中国香港爆发的禽流感，1998年东南亚猪脑炎事件，1999年比利时等国二噁英事件，2001年欧洲爆发的口蹄疫，2008年中国令人震惊的三鹿乳粉事件，

2009年全球性的甲型H1N1流感,2009年中国的"瘦肉精"事件,2011年发生在中国的地沟油事件等。这些食品安全问题不仅危害了消费者的身心健康,而且给国家的经济造成了不可弥补的损失。据国家卫生部公布的全国食物中毒情况,仅2003年,全国重大食物中毒事件达379起,中毒12,876人,死亡323人;2007年,全国发生重大食物中毒事件506起,中毒13,280人,死亡258人;2008年,全国发生重大食物中毒事件437起,中毒13,325人,死亡152人。

食品安全问题也从侧面反映了当今的食品安全存在着诸多隐患,分析其原因主要有:食品加工过程中使用劣质原料,添加有毒物质,超量使用食品添加剂,滥用非食品加工用化学添加剂。农产品、禽类产品的安全状况也不容乐观,抗生素、激素和其他有害物质残留于禽、畜、水产品体内。转基因食品对人类也有潜在威胁,尽管目前还没有足够证据证明转基因食品对人类有害。当然,出现食品安全问题也有社会自身的原因,如商家为了牟取暴利,不顾食品安全及产品质量;企业管理混乱;新技术不断出现带来新的食品质量安全问题;社会监管制度以及法律不健全等。

目前,如何给消费者提供安全的食品仍然是一些发达国家以及发展中国家所面临的一大挑战,也是世界各国决策者关注的焦点。要解决食品安全问题,就需要清楚可能出现食品安全问题的环节以及影响食品安全的因素,最后通过制定法规和加强监督来减少甚至杜绝食品安全问题的发生。本章主要分析整个食物供应链可能出现的食品安全问题,其中穿插介绍近几年世界各地发生的重大食品安全事件,目的是使人们了解从农田到餐桌食品安全问题可能产生的环节,最后针对食品供应链出现的食品安全问题提出一些应对措施。

2 食品安全问题来源于整个食物供应链

食品安全问题源于整个食物供应链,原料生长的环境,气候和土壤,加工,包装,贮藏,运输以及销售等环节都有可能出现食品安全问题。图1概括了在食品供应链中可能出现的食品安全问题及风险。这些问题和风险主要来源于人类的行为,人类行为表现出较强的复杂性和不确定性。也与产地,食品原料,食品的类型,属性以及加工技术和过程等因素有直接的关系。

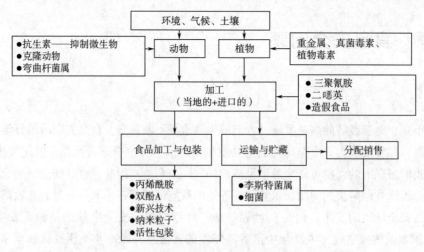

图1 食品供应链中可能出现的食品安全问题及风险

2.1 环境、气候和土壤

食品安全问题产生的源头是食品加工原料生长的环境、气候和土壤。影响环境、气候和土壤因素主要是自然灾害、污染和气候。图2是环境、气候和土壤对食品污染的影响。

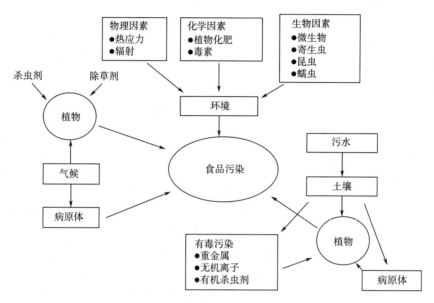

图2 环境、气候和土壤对食品污染的影响

2.1.1 环境

人类对环境及生物圈的污染已经对全球的生命构成了持续威胁,各种各样的环境污染造成食品污染,最后食品污染严重损害了人类DNA遗传物质。人类对环境的污染及环境对人类的威胁是同时产生的。生活垃圾、动物饲料、植物化肥是环境污染和食品污染的良好携带者。影响环境污染的因素包括物理因素,如热应力(热浪和寒潮)和辐射等;化学因素,如植物化肥和有毒物质等;生物因素,如微生物、寄生虫、昆虫和蠕虫等。而环境污染反过来又会影响大气、土壤、水、动物、植物和微生物,最终也会导致食品安全问题。

2.1.2 气候

气候变化对食品安全的潜在影响一直在调查和争论中。气候变化会影响植物产品在贮藏过程中真菌毒素的形成。气候变化造成了大量昆虫的繁殖从而影响植物作物的生长,气候变化也影响田间杂草的生长。农药以及除草剂的使用导致了农作物中农药及有害化学物的残留问题。病原菌在极端天气条件下也会大量繁殖,如洪水和热浪。气候变化的结果是改变了整个食品体系,包括从农场到餐桌的所有阶段(主要是初级生产、加工、贮运和贸易)。

2.1.3 土壤

农作物的生长离不开土壤,土壤质量的好坏直接关系到食品原料的安全问题。目前,生活污水和未经处理的工业废水是造成土壤污染的主要原因。

2.1.3.1 病原体污染物

生活污水、畜禽饲养场污水以及制革、屠宰业和医院等排出的废水,常含有各种病原体,如病毒、病菌、寄生虫。这些病原体保留在土壤中，又从土壤转移到农作物，从而造成食品安全问题。受病原体污染后的土壤中微生物激增，其中许多是致病菌、病虫卵和病毒，它们往往与其他细菌和大肠杆菌共存，病原体污染物可通过多种途径从土壤进入农作物，一旦条件适合，就会引起人体疾病。

2.1.3.2 有毒污染物

有毒污染物指的是进入生物体后累积到一定数量能使体液和组织发生生化和生理功能的变化，引起暂时或持久的病理状态，甚至危及生命的物质。如重金属和难分解的有机污染物等。污染物的毒性与摄入机体内的数量有密切关系。由于生活污水及工业废水的排放造成的土壤有毒污染物主要有：①重金属。重金属被看作是土壤污染中最严重的问题，农作物能够从污染的土壤及大气中吸收重金属并且可以经过长期富集，富集大量重金属的农作物会导致很严重的食品安全问题。重金属如镉和铅已被证明有致癌的作用，水果和蔬菜中含有高浓度的铜，镉和铅导致上消化道癌发病率很高。重金属在自然界中一般不易消失，它们能通过食物链而被富集。重金属优先积累在人体组织中，例如，砷、铬、镉、铅会影响神经系统和肾脏，以及造血系统（血液）。镉和铬对肝脏有影响，它们会导致男性生殖系统的损害。这类物质除直接作用于人体引起疾病外，某些重金属还可能促进慢性病的发展。②无机阴离子，由一些致癌物组成，如NO^-、F^-、CN^-。剧毒物质氰化物主要来自工业废水排放。土壤中的这些无机阴离子会直接或间接地作用于农作物，从而导致食品安全问题。③有机农药和多氯联苯。目前世界上有机农药大约6000种，常用的有200多种。农药喷在农田中，经淋溶等作用进入土壤，造成污染。有机农药可分为有机磷农药和有机氯农药。有机磷农药的毒性虽大，但一般容易降解，积累性不强，因而对生态系统的影响不明显；而绝大多数的有机氯农药，毒性大，几乎不降解，积累性甚高，对生态系统有显著影响。农作物也能够从污染的土壤中吸收有机农药导致农药残留并且可以经过长期富集而导致很严重的食品安全问题。

2.2 动物和植物

在环境、气候和土壤条件下生长的动物和植物，是食品加工的原料。环境、气候和土壤条件直接关系到这些食品加工原料的好坏，同时关系到食品安全问题。对于植物而言，一些化学物质如重金属、无机阴离子及有机农药的残留等，一些生物物质如真菌毒素类和藻类毒素都会导致食品安全问题的发生。而影响动物的安全因素有抗生素的使用，克隆动物以及寄生于哺乳动物和禽类肠道里的致病弯曲杆菌属等。

2.3 食品加工及包装

食品原料在加工过程中会出现较频繁的食品安全问题，如添加三聚氰胺、二噁英以及利用劣质原料加工食品等。食品加工过程中出现的食品安全问题，包括污染物的不完全物理去除、微生物、孢子、病毒、寄生虫（对冷藏的动物食品危害比较大）的不彻底灭活、天然的抗营养因子、过敏原及有毒物质的不彻底灭活。还有由于过度加工导致营养物质的损失，毒性分子的形成（丙烯酰胺、杂环胺和5-羟甲基糠醛），促进微生物的生长（如切割、磨细、长期贮藏）。在食品包装过程中，包装材料中有害化学物质从包装向食品的转移。以下是在食品加工及包装过程中可能出现食品安全问题的主要物质。

2.3.1 食品加工及包装存在的有毒物质
2.3.1.1 丙烯酰胺

丙烯酰胺（分子式 C_3H_5NO）主要形成于富含碳水化合物的食物中，如马铃薯和谷类食品（薯片、炸薯条、面包、脆面包、饼干、谷类早餐食品及咖啡）。这些食品可能在烹饪过程中形成丙烯酰胺，如煎炸，烘烤和微波炉加热（>120℃）。在薯片及饼干中丙烯酰胺含量分别达到 5mg/kg 和 2mg/kg。在加热的富含脂肪的干燥食品如坚果中也会形成丙烯酰胺。虽然丙烯酰胺形成的机理目前还不清楚，但是它涉及天门冬酰胺和还原糖之间的美拉德反应。丙烯酰胺也是一种高活性的工业化学物质，应用于塑料和原料的生产。它可以从食品接触的原料中迁移进入食品。动物实验表明丙烯酰胺能致癌，或许是一种生殖毒物。尽管人类接触到丙烯酰胺的食物有很长一段时间，但丙烯酰胺与人类癌症风险之间的联系目前还不确定。近年的研究表明致癌物丙烯酰胺的日允许摄取量规定在 2.6mg/kg 体重，加拿大、瑞典以及美国研究认为成年人丙烯酰胺每天的平均暴露量分别规定在 0.3，0.5，0.4mg/kg 体重。

2.3.1.2 双酚 A

双酚 A（BPA）又称 4,4-二羟基-2,2-二苯基丙烷，该物质经常和其他化学物质结合使用生产塑料和树脂。因此双酚 A 存在于聚碳酸酯食品容器中，如可回收饮料瓶、婴幼儿奶瓶、餐具和贮藏容器中。双酚 A 残留物也存在于环氧树脂中，该环氧树脂用来制作食品和饮料罐头和大桶的保护涂层。食品在贮藏过程中，贮藏容器中的双酚 A 有一小部分可以迁移进入食品和饮料中。双酚 A 对女性生殖器有副作用。关于双酚 A 的安全性，欧洲食品安全局（EFSA）规定的平均日摄取量（TDI）为 50 μg/kg 体重，新生儿、儿童和成年人的平均日摄取量要低于该值。关于双酚 A 各种各样的毒副作用仍然存在争议。在美国、欧盟国家以及日本，双酚 A 是允许用在食品容器中，但是美国和欧盟国家禁止双酚 A 用于婴幼儿食品容器中。

2.3.2 伪劣和掺假食品

有些企业为了牟利，通过直接或间接向食品中添加有毒物质导致了大量的食品安全问题。以下是一些伪劣和掺假食品安全问题。

2.3.2.1 劣质和掺假油

乌克兰将掺有矿物油的葵花籽油出口到欧盟，最后导致从欧盟市场撤回。西班牙毒油事件导致 20,000 人被感染以及 800 多人死亡，是由于商家将一种有毒的改性菜籽油冒充食用橄榄油进行销售。2011 年，中国发生了地沟油事件。地沟油泛指在生活中存在的各类劣质油，如回收的食用油、反复使用的炸油等。地沟油最大来源为城市大型饭店下水道的隔油池。长期食用可能会引发癌症，对人体的危害极大。其主要成分仍然是甘油三酯，却又比真正的食用油多了许多致病、致癌的毒性物质。

2.3.2.2 三聚氰胺

2008 年，中国的三聚氰胺事件，大约有 30 万婴幼儿和儿童食用了含有三聚氰胺的婴幼儿乳粉，导致了肾脏和泌尿道系统疾病，包括肾结石，据报道有 6 个死亡病例。调查表明其主要原因是将三聚氰胺添加到稀释牛乳中以提高稀释乳的蛋白质含量。由于三聚氰胺便宜易得，商家在经济利益的驱使下非法添加。三聚氰胺分子式为 $C_3H_6N_6$，其氮含量非常高，占整个质量的 66%，但是三聚氰胺是非蛋白质类分子，因此它的添加只能导致一种高蛋白的假象。三聚氰胺的急性毒性低，小鼠和大鼠的口服半数致死剂量是 3,000mg/kg 体重，在剂量上升到

18,000mg/kg 体重时，被发现毒性会影响排泄器官、肾和膀胱，而且这些影响在男性器官中比女性更明显。

2.3.2.3 二噁英

二噁英是一种无色无味、毒性严重的脂溶性物质。2008年，欧盟食品和饲料类快速预警系统（RASFF）通报指出食品中的二噁英超出了法定限量。爱尔兰的猪肉及猪肉产品由于二噁英超标从54个国家被召回。欧盟食品和饲料类快速预警系统年报报道，该猪肉中的二噁英已经超出了欧盟最大限量1pg/g脂肪的100倍。污染或掺假的源头后来被调查，来源于50个农场的猪被饲喂了污染的焙烤废弃物，这些废弃物是由被燃烧气体直接烘干的面包碎屑组成，产生该燃烧气体的燃油由于非法添加了可以产生二噁英的油从而导致面包碎屑受污染。

2.3.3 食品添加剂

食品添加剂的种类有色素、防腐剂、抗氧化剂、乳化剂、稳定剂和增甜剂等。一些食品添加剂有神经毒性，意味着它们能够改变神经系统的正常生理活动，甚至能够杀死神经元。目前，非法使用食品添加剂生产食品已经不是个例。2010年，一些中国小麦粉里掺入添加漂白剂的石灰粉。

2.3.4 伪劣、假冒食品和饮料

在食品和饮料工业中，据2009年统计，全球的假冒食品估计大约在每年500亿美元。假冒食品和饮料出现频率较高的是：水果、保鲜蔬菜、婴幼儿食品、乳粉、黄油、速溶咖啡、烈性酒、饮料和糖果。很多假冒食品和饮料通过冒充名牌产品来获得高的利润。由于假冒产品的质量达不到要求而造成食品安全问题。

2.3.5 新兴食品加工技术

一些新兴的食品加工技术也会对食品安全造成潜在的威胁，而有些技术也广泛应用于食品工业。对一些新型技术可能造成的食品安全问题简单介绍如下。

2.3.5.1 新型物理处理技术

高压杀菌：微生物和一些孢子没有完全灭活、发生某些化学反应；

脉冲电场：孢子没有完全灭活、发生电化学反应、金属从电极发生迁移；

冷等离子体：产生自由基、发生氧化；

光脉冲与紫外：发生光氧化反应；

超声：微生物和一些孢子没有完全灭活；

欧姆加热：金属从电极发生迁移；

电离辐射：产生游离基、发生氧化反应。

2.3.5.2 新型化学处理技术

气调：能延缓微生物的生长，但是保质期及贮藏时间过长；

抗菌剂：微生物和一些孢子没有完全灭活，能延缓微生物的生长，有潜在毒性。

2.3.6 纳米技术

工程纳米材料或纳米微粒已经广泛用于食品成分、添加剂、化肥、农药、药物及包装材料中。工程纳米材料有着特殊的物化特性。纳米技术在食品中的用途主要有如下方面。

（1）作为食品成分及添加剂　增加可溶性、分散性，在没有乳化剂的情况下稳定乳化液，

提高产品质地;

(2) 作为生物活性物质的输送体系　保护目标营养物的释放,提高生物活性物质的利用率;

(3) 新型包装材料　抗菌膜。

纳米技术在食品工业中的应用引起的食品安全问题还在讨论和研究之中。如纳米材料的检测、性质及剂量问题,纳米材料的吸收、代谢及分泌问题,纳米材料的毒性及对环境的影响,都有待进行安全评估及授权。

2.3.7　其他挑战

食品安全问题正面临着技术创新需求与食品安全问题之间的矛盾。虽然有欧洲食品安全局(EFSA)和食品法规的监管、更严格的安全和质量标准以及强化质量监控,但是食品安全事故不断增加,消费者对于食品安全的信心不断下降。一些农业食品技术往往被消费者拒绝:

离子辐射食品;

激素(和抗生素)治疗动物和加速动物的生长,增加肉产品和乳产品产量(欧盟国家禁止);

各种各样的食品添加剂;

肥料及农药的过量使用;

转基因食品及食品成分;

转基因动物(包括克隆动物)。

在新技术新产品不断出现的同时,也给各种风险评估方法如病原菌的监测、过敏原检测、毒理评价及对环境的影响造成了较大的挑战。

在欧盟国家,消费者对产品的知情选择权已经建立。辐照食品和转基因食品的强制性标签阻碍了制造商和零售商将这些产品推向市场,尽管它们有潜在的优势。

2.3.8　食品活性包装

食品活性包装指通过检测包装食品的环境条件,提供在流通和储存期间包装食品品质的信息。如时间-温度显示、新鲜度显示、包装泄漏显示等。食品活性包装给食品安全问题带来的隐患在于包装材料中的化学物质及其降解产物会迁移进入食品中。纳米粒子也能造成食品安全风险,欧洲食品安全局(EFSA)对于食品活性包装技术有授权计划,重点关注化学物质的迁移及其毒理学特性。

2.4　食品运输和贮藏

食品在运输和贮藏过程中,如果冷藏分配渠道管理不当,就会被微生物污染。对于一些即食食品(pH>4.5或水分活度>0.91)很容易导致单核细胞增生李斯特菌的大量繁殖。一些重大疫情与新鲜蔬菜和水果有关,如沙拉用的绿叶蔬菜、莴苣、新芽和甜瓜,其中最常见的病原体是:诺如病毒、沙门氏菌、大肠杆菌O157。

2.5　食品的分配和销售

食品的分配和销售和食品的运输和贮藏一样,如果处理不当,也会导致微生物污染。

3 针对食品安全问题的应对措施

食品安全问题可能出现在食品供应链的任何一个环节，如果任何一个细小的环节监控和管理不当，都会出现食品安全问题。因此，控制和减少食品安全问题的发生对于全世界来说都是一个很大的挑战。针对在食品供应链中可能出现的问题，提出一些相应的应对措施。

3.1 建立合理的食品安全管理工具

合理的食品安全管理工具对于避免食品安全问题的发生起到很关键的作用。这些工具包括食品立法、建立国际和国内食品标准、质量管理体系、风险评估分析、风险监测和控制体系、建立食品污染的监测和预警系统、微生物风险评估、营养和毒理学研究体系、精细的食品加工技术。这些工具让消费者对安全食品的选择有知情权。

3.2 建立预测、预警和通信系统

建立预测、预警和通信系统可以使食品危害在成为真正的风险之前及时采取预防措施。

3.3 监管食品供应链的每个环节

食品原料加工、运输、贮藏及分配都要尽可能避免生物和化学污染。近几年的食品安全事件表明，二噁英、丙烯酰胺、三聚氰胺和双酚A等化学有害物质的污染以及大量病原菌的爆发仍然是导致食品安全问题频繁发生的主要原因。因此，要减少和杜绝食品安全问题的发生就需要严格监管食品供应链的每个环节。

3.4 建立严格的市场监管制度

三聚氰胺事件，国际国内的假冒食品和饮料以及掺假食品对食品安全的威胁不断增加，这些事件的发生都是利益驱使下的欺诈行为，需要严格的市场监管制度进行监管。

3.5 对新技术的应用建立严格的上市前的安全评估和授权程序

食品在加工过程中会用到一些新技术，新技术的应用对食品安全问题的影响还存在争论，需要对一些新技术进行严格的安全评估和授权程序，让消费者解除顾虑，选择安全放心的食品。

LESSON 4

FOOD PRESERVATIVES

1 FOOD ADDITIVES

Food additives are defined as a class of substances that are used in the production, processing, storage and packaging of food so as to extend the shelf-life and meet the processing needs for improving the food quality, color, flavor and taste. The use of food additives should meet three basic conditions. Firstly, it is necessary. Secondly, it is safe or reliable and should be proved to be harmless to human body by toxicological experiments. Thirdly, the government's permission and strict regulation on the variety, scope, and dosage of food additives that are allowed to be used have been established in our country. According to the differences in their functions, food additives can be divided into 23 groups, which mainly include acidity regulator, anticaking agent, antifoaming agent, antioxidant, bleaching agent, raising agent, rubber base in candy based agent, colorant, color-protecting, emulsifiers, enzyme preparation, flavor enhancer, flour treatment agent, coating agent, furthermore, water retention agent, nutrition enhancer, preservatives, stabilizers, curing agent, sweeteners, thickening agent, food spices, and food processing aids used in industry and others. Food additives can maintain and improve the nutritional value of food; and can play an important role in maintaining food safety and extending the shelf life of food. It can meet the needs of different people and the requirement of food processing, such as lubrication, defoaming, filtration, stabilization and coagulation. The legitimate use of food additives is safe and necessary. With the development of the food industry and the improvement of people's living standards, food additives, with their increasing variety and quantity, have become an indispensable part of the production of the food industry.

2 DEFINITION AND CLASSIFICATION OF FOOD PRESERVATIVES

2.1 Definition of Food Preservatives

Food preservatives are a series of substances added to food to prevent decomposition and prolong the shelf life of the food. More specifically, preservatives refer to a class of substances used in food processing to improve the food preservation and retain the edibleness of food by inhibiting the microbial activity and preventing spoilage caused by microorganism contamination in the process of production, transportation, storage and circulation. They generally play the functions of antibacterial and antioxidant agents, which can not only inhibit the growth of molds, yeasts and bacteria, but also prevent food from producing rancid odor and restrain the formation of browning and dark spots.

Characteristics of Food Preservatives:

Stable property, being effective within a certain period of time;

Non-toxic during usage or after decomposition, and neither hinder the normal function of gastrointestinal enzymes, nor affect the intestinal normal probiotics;

Bacteriostatic or bactericidal effect at low concentrations;

No pungent and abnormal smell;

Easy to use, with a reasonable price.

2.2 Classification of Food Preservatives

According to different sources, food preservatives are usually divided into two categories: synthetic preservatives and natural preservatives. Most of the preservatives existing in the market are artificially synthesized chemical preservatives. Chemical food preservatives can be divided into three categories: acid preservatives, ester preservatives, and inorganic salt preservatives. Acid preservatives mainly include benzoic acid, sorbic acid, propionic acid and their salts; the characteristic of acid preservatives is that, the greater acidity of the system, the better the preservative effect, but almost ineffective under alkaline conditions. Ester preservatives include parabens, gallates, ascorbyl palmitate and so on, featuring low toxicity and an effectiveness over a wide pH range. Inorganic salt preservatives include sulfurous compounds like sulfites, pyrosulfites, nitrates and nitrites, which are characterized by being not only antimicrobially active, but being able to inhibit discoloration by blocking compounds with a reactive carbonyl group (maillard reaction; nonenzymatic browning) or by inhibiting oxidation of phenols by phenol oxidase enzymes (enzymatic browning).

Natural food preservatives are a class of antiseptic substances separated and extracted directly from plants, animals, or microorganisms, also known as biological preservatives. According to different sources, natural food preservatives can be divided into three categories: animal-derived, plant-derived and microbial-derived. Animal-derived natural preservatives refer to the preservatives extracted from animals, which commonly include propolis, protamine and chitosan etc.; its characteristics are highly nutritional, natural, non-toxic, but scarce and expensive. Plant-derived natural preservatives refer to the effective antibacterial components extracted from plants, such as tea polyphenol, clove oil, allitridin and so on; their characteristics are the narrow antibacterial spectrum of microorganisms and have a bad influence on the color and taste of product. Microbial-derived preservatives refer to the bacteriostatic substances isolated and extracted by microorganisms, mainly including nisin and natamycin, etc., characterized by being safe, high efficient, and non-toxic.

3 ACTION MECHANISM OF FOOD PRESERVATIVES

Food deterioration refers to the change process of food quality (physical and chemical properties) under the influence of some certain factors (intrinsic and extrinsic). From harvesting, slaughtering and manufacturing, the quality of food begins to change under the influence of environmental conditions, mostly change to unfavorable directions. After deterioration, the edible value of food will reduce, and may harm human health after being eaten.

The main causes of food deterioration are listed as follows:

(1) Oxidation and deterioration of food ingredients caused by air oxidation and drying effect, results in rancidity, loss of vitamins and linked browning. In the high temperature seasons, the dry air makes food lose its freshness and texture.

(2) Deterioration caused by microbial contamination and proliferation, such as the spoilage of food caused by the protein degradation, oxidative rancidity of carbohydrate or fats due to the microbial action and so on.

(3) Decomposition of food caused by the endogenous enzymes such as oxidase, amylase and protease, which may generate thermal energy, water vapor and carbon dioxide, and gradually results in the deterioration of food.

(4) Spoilage of food caused by the erosion and reproduction of insects, and direct or indirect contamination of harmful substances.

The mechanism of food preservatives to inhibit and kill microorganisms is very complex. It is generally believed that the mechanism of current food preservatives on microorganisms includes the following aspects:

(1) Acting on the cell wall and cell membrane system, by disrupting the structure of microbial cell membrane or changing the permeability of the cell membrane to achieve bacteriostatic or bactericidal purposes, such as acidic preservatives.

(2) Acting on enzymes or functional proteins, for example, by utilizing covalent bonds formed by the conjugation between the own double bonds and the enzymatic sulfydryl of microbial cells, sorbic acid to deprive the enzyme of its activity and to disrupt many functions of important enzyme systems.

(3) Targeting genetic material or genetic microstructure, for example, lysozyme can directly combine with negatively charged viral proteins, triggering the virus inactivation through the formation of compound salt bound with DNA, RNA and apoprotein.

4 ATTENTIONS FOR THE USE OF FOOD PRESERVATIVES

Use of food preservatives must strictly observe the "Standards for uses of food additives" (GB 2760-2014).

For rational usage, food preservatives should be selected correctly, and the following points should be considered:

(1) Understand the physical and chemical properties of food preservatives, such as solubility, heat resistance, optimal pH, antibacterial spectrum and minimum inhibitory concentration.

(2) Understand the quality of food itself and the condition of the bacteria.

(3) Understand the environmental conditions in food processing, storage and transportation, so as to ensure the best preservative performance of food preservatives.

When using food preservatives, the following principles should be observed: (1) Rational use, harmless to human health; (2) No effect on the gastrointestinal flora; (3) Degradable to the normal components of food in the digestive tract; (4) No effect on the use of antibiotics; (5) No harmful ingredients produced during heat treatment.

In the production practice, the scientific and rational use of food preservatives can be carried out in the following aspects:

(1) Before adding antiseptic, it should be ensured that the food has been sterilized completely, and there is no large number of microorganisms in order to reach desirable effect when adding preservatives. For example, potassium sorbate won't play a preservative effect when there is a large amount of microorganisms, but will become a nutrient source of microbial reproduction.

(2) Need to know the toxicity of various types of preservatives and the range of their application. The scope and quantity preservatives shall be in line with the state standards on the premise of harmlessness to human body.

(3) Need to know the kinds of microorganisms that preservatives can inhibit. Some preservatives have effect on mold, and some have effect on yeast. Generally, compound antiseptics are applied in a comprehensive preservation process.

(4) Need to know the effective use environment of the various types of preservatives. For example, the antiseptic effect of acidic preservatives is greatly influenced by food pH. The lower the pH, the better the effect. Thus, compound preservatives are a good recommendation.

(5) According to different kinds of food processing technology, the price and solubility of preservatives should be considered, as well as affecting factors on food flavor preservatives should be added with flexibility on the basis of these distinctive features.

(6) Need to know the physical and chemical properties, processing, packaging, storage conditions of foods and their effects on preservatives, and determine the timing of preservatives addition.

Lesson 4

Vocabulary

Words

toxicological [ˌtɑksəkəˈlɑdʒɪkəl] adj. 毒理学的
antioxidant [ˌæntiˈɑːksɪdənt] n. 抗氧化剂
emulsifier [ɪˈmʌlsɪfaɪər] n. 乳化剂
preservative [prɪˈzɜːrvətɪv] n. 防腐剂
stabilizer [sˈteɪbəlaɪzər] n. 稳定剂
sweetener [ˈswiːtnər] n. 甜味剂
spoilage [ˈspɔɪlɪdʒ] n. 腐败
bacteriostatic [bækˌtɪriəsˈtætɪk] adj. 抑菌的
paraben [pəˈræben] n. 对羟基苯甲酸酯
pyrosulfite [paɪroʊˈsʌlfaɪt] n. 焦亚硫酸盐
protamine [ˈproʊtəmiːn] n. （鱼）精蛋白
nisin [ˈnaɪsɪn] n. 乳酸链球菌素
lubrication [ˌluːbrɪˈkeɪʃn] n. 润滑
defoaming [dɪˈfoʊmɪŋ] n. 消泡
filtration [fɪlˈtreɪʃn] n. 过滤
stabilization [ˌsteɪbəlaɪˈzeɪʃn] n. 稳定
coagulation [koʊˌægjuˈleɪʃn] n. 凝结
colorant [ˈkʌlərənt] n. 着色剂
gastrointestinal [ˌgæstroʊɪnˈtestɪnl] adj. 胃肠道的
bactericidal [bækˌtɪrɪˈsaɪdl] adj. 杀菌的
gallate [ˈgæleɪt] n. 没食子酸盐
palmitate [ˈpælməteɪt] n. 棕榈酸盐
propolis [ˈprɒpəlɪs] n. 蜂胶
natamycin [neɪtəˈmaɪsɪn] n. 纳他霉素

Phrases

acidity regulator 酸度调节剂
anticaking agent 抗结剂
antifoaming agent 消泡剂
bleaching agent 漂白剂
raising agent 膨松剂
rubber base in candy based agent 胶基糖果中基础剂物质
enzyme preparation 酶制剂
food preservatives 食品防腐剂
benzoic acid 苯甲酸
propionic acid 丙酸
gastrointestinal flora 胃肠菌群
food deterioration 食品变质
flavor enhancer 风味增强剂
flour treatment agent 面粉处理剂
water retention agent 保水剂
nutrition enhancer 营养强化剂
curing agent 凝固剂
texturizing agent 增稠剂
color-protecting 护色剂
food spices 食用香料
rancid odour 腐臭气味
sorbic acid 山梨酸
tea polyphenol 茶多酚
potassium sorbate 山梨酸钾
endogenous enzyme 内源性酶

Exercises

I. Write true or false for the following statements according to the passage.

1. ( ) Bacteria, yeasts, and molds are the main factors that cause food spoilage.

2. (　　) All food preservatives are synthetic and harmful.

3. (　　) All food preservatives can be used under alkaline conditions.

4. (　　) Natural preservatives are generally non-toxic and harmless.

5. (　　) Food preservatives can inhibit microorganisms by acting on microbial genetic material or genetic microstructure to make it dead.

Ⅱ. **Multiple choice questions**（Choose the correct answers according to the passage）.

1. Which of the followings are natural preservatives? (　　)

A. potassium sorbate　　　　　　B. sodium diacetate

C. nisin　　　　　　　　　　　　D. tea polyphenols

2. Factors that affect the antiseptic effect of preservatives (　　)

A. the pH of the food system　　　B. food contamination level

C. kinds of microorganisms　　　　D. types of preservatives

3. Ester preservatives include (　　)

A. benzoic acid　　　　　　　　　B. parabens

C. nitrate and nitrite　　　　　　　D. gallates

Ⅲ. **Answer the following questions according to the passage.**

1. What is the definition of food preservatives?

2. What are the characteristics of food preservatives?

3. How to choose food preservatives correctly?

4. What are the main causes of food deterioration?

5. How to use food preservatives correctly?

Ⅳ. **Translate the following words and expressions into Chinese.**

acidic preservatives　　　　　　　synthetic preservatives

enzymatic browning　　　　　　　bacteriostatic substances

bactericidal effect　　　　　　　　antibacterial spectrum

inorganic salt preservatives　　　　compound antiseptics

potassium sorbate　　　　　　　　sulfurous compounds

Ⅴ. **Translate the following expressions into English.**

1. 酸性防腐剂主要包括苯甲酸、山梨酸和丙酸以及它们的盐类，其特点是体系酸性越大，其防腐效果越好，但是在碱性条件下几乎无效。

2. 天然食品防腐剂是从植物、动物、微生物中直接分离提取的具有防腐作用的一类物质，也称作生物防腐剂。根据来源可分为动物源、植物源和微生物源天然防腐剂三大类。

3. 食品的变质是指在某些因素（内在、外在）的影响下，食品质量（理化性质）发生变化的过程。

参考译文

第4课 食品防腐剂

1 食品添加剂

食品添加剂是指为改善食品品质和色、香、味，满足加工需要，延长保质期等目的，在食品加工贮藏、生产包装等过程所用到的一类物质。食品添加剂的使用需满足三个最基本的条件：一是确有必要；二是安全可靠，使用的食品添加剂必须是经过毒理学实验证明对人体无害的；三是政府许可，我国对允许使用的食品添加剂的品种、范围和用量都有明确的规定。按功能的不同，将食品添加剂分为23类，主要包括酸度调节剂、抗结剂、消泡剂、抗氧化剂、漂白剂、膨松剂、胶基糖果中基础剂、着色剂、护色剂、乳化剂、酶制剂、增味剂、面粉处理剂、被膜剂、水分保持剂、营养强化剂、防腐剂、稳定剂、凝固剂、甜味剂、增稠剂、食品用香料、食品工业用加工助剂及其他。食品添加剂可以保持和提高食品的营养价值，对保证食品安全、延长食品的保质期具有重要作用。能够满足食品加工操作的需要，如食品加工过程中许多需要润滑、消泡、助滤、稳定和凝固等，同时可以满足不同人群的需要。合法使用食品添加剂不仅是安全的，也是必要的。随着食品工业发展，人民生活水平提高，使用食品添加剂的品种和数量越来越多，已成为现代食品工业生产中不可缺少的物质。

2 食品防腐剂的定义及分类

2.1 食品防腐剂的定义

食品防腐剂是一类加入食品中能防止食品腐败变质、延长食品保质期的物质。具体而言，指为了抑制微生物活动，防止食品在生产、运输、贮藏、流通过程中因微生物繁殖污染而引起的腐败变质以及提高食品的保存性、延长食用价值，而在食品加工过程中添加的一类物质。它一般具有抗菌剂和抗氧化剂的双重作用，既可以抑制霉菌、酵母、细菌的生长而起到防腐作用，还可以使食物不产生腐臭，抑制褐变和黑斑的形成。

食品防腐剂具有以下特征：

性质稳定，在一定的时间内有效；

使用过程中或分解后无毒，不阻碍胃肠道酶类的正常作用，也不影响肠道有益的正常菌群的活动；

在较低浓度下有抑菌或杀菌作用；

本身无刺激性气味和异味；

使用方便，价格合理。

2.2 食品防腐剂的分类

依据不同来源，防腐剂可以分为合成防腐剂和天然防腐剂两大类。目前市场上使用的防腐剂绝大多数是化学防腐剂，化学类食品防腐剂可分为三大类：酸性防腐剂、酯类防腐剂、无机盐类防腐剂。酸性防腐剂主要包括苯甲酸、山梨酸和丙酸以及它们的盐类，其特点是体系酸性越大，其防腐效果越好，但是在碱性条件下几乎无效。酯类防腐剂主要包括对羟基苯甲酸酯类、没食子酸酯、抗坏血酸棕榈酸酯等，其特点是在很宽的 pH 范围内都有效，毒性比较低。无机盐类防腐剂主要包括含硫的亚硫酸盐、焦亚硫酸盐、硝酸盐及亚硝酸盐等，其特点是不仅具有抗菌活性，而且还能通过阻止活性羰基化合物抑制变色反应（美拉德反应；非酶褐变）或通过对酚氧化酶的作用抑制酚的氧化（酶促褐变）。

天然食品防腐剂是从植物、动物、微生物中直接分离提取的具有防腐作用的一类物质，也称作生物防腐剂。根据来源可分为动物源、植物源和微生物源天然防腐剂三大类。动物源天然防腐剂指从动物体内提取出来的防腐剂，常用的主要有蜂胶、鱼精蛋白、壳聚糖等，其特点是本身就是含较高营养价值的食品，纯天然、无毒、资源较为稀缺、价格高。植物源天然防腐剂指从植物中提取出来的有效抑菌成分，如茶多酚、丁香油、大蒜素等，其特点是抑菌面窄，并且会影响产品色泽和味道。微生物源防腐剂指由微生物产生而分离提取的抑菌物质。主要有乳酸链球菌素和纳他霉素等，具有安全、高效、无毒的特点。

3 食品防腐剂的作用机理

食品的变质是指在某些因素（内在、外在）的影响下，食品质量（理化性质）发生变化的过程。食品从收获、屠宰、制造起，受环境条件的影响，质量就开始变化，绝大多数是质量向不利的方向变化。变质的食物，食用价值下降，食用后可能危害人体健康。

食品产生变质的原因主要有：

（1）因空气的氧化与干燥作用引起食品成分的氧化变质，导致油脂酸败、维生素的损失及连锁褐变；高温季节里，空气的脱水作用使食品丧失了新鲜和充盈的质感；

（2）因微生物的污染、繁殖引起的变质，其中有由于食品中蛋白质被微生物分解造成的腐败，食品中碳水化合物或脂肪被微生物分解产酸而产生的酸败等；

（3）因食品内部所含氧化酶、淀粉酶、蛋白酶的作用使食品分解，产生热能、水蒸气和二氧化碳，致使食品逐渐变质；

（4）因昆虫的侵蚀、繁殖和有害物质的直接或间接污染导致食品的腐败变质。

食品防腐剂抑制和杀死微生物的机理十分复杂，一般认为，当前使用的食品防腐剂对微生物的作用机理包括以下几个方面：

（1）作用于细胞壁和细胞膜系统，通过破坏微生物细胞膜的结构或者改变细胞膜的渗透性达到抑菌或杀菌目的。如酸性防腐剂；

（2）作用于酶或功能蛋白，如山梨酸利用自身的双键与微生物细胞中的酶的巯基结合形成共价键，使其丧失活性，从而破坏许多重要酶系统的功能；

(3) 作用于遗传物质或遗传微粒结构，如溶菌酶可与带负电荷的病毒蛋白直接结合，与DNA、RNA、脱辅基蛋白形成复盐，使病毒失活。

4 食品防腐剂的使用注意事项

食品防腐剂的使用要严格遵守《食品安全国家标准　食品添加剂使用标准》（GB 2760—2014），在合理使用食品防腐剂前，先要正确选用食品防腐剂，可以从以下几点考虑：（1）了解食品防腐剂的理化特性，如溶解性、耐热性、最适pH、抗菌谱和最低抑菌浓度等；（2）了解食品本身的品质和染菌情况；（3）了解食品加工、贮藏和运输过程中的环境条件，确保食品防腐剂发挥最佳防腐性能。

使用食品防腐剂时，应遵循的原则是：（1）合理使用，保证对人体健康无害；（2）不影响消化道菌群；（3）在消化道内可降解为食物的正常成分；（4）不影响药物抗生素的使用；（5）食品热处理时，不产生有害成分。

在生产实践中，食品防腐剂的科学、合理使用，可以从以下几个方面着手。

（1）在添加防腐剂之前，应保证食品灭菌完全，不允许存在大量的微生物，否则防腐剂的加入将不会起到理想的效果。如山梨酸钾，在微生物大量存在时不但不会起到防腐的作用，反而会成为微生物繁殖的营养源。

（2）应了解各类防腐剂的毒性和使用范围，食品防腐剂的使用范围和使用量应当按照国家标准的规定执行，以对人体无毒害为前提。

（3）应了解各类防腐剂所能抑制的微生物种类，有些防腐剂对霉菌有效果，有些对酵母有效果。一般以复配防腐剂进行综合防腐保鲜处理。

（4）应了解各类防腐剂的有效使用环境。例如，酸性防腐剂的杀菌效果受食品pH的影响较大，pH越低，效果越好。因此，建议采用复合防腐剂。

（5）根据各类食品加工工艺的不同，应考虑防腐剂的价格和溶解性，以及对食品风味是否有影响等因素，综合其优缺点灵活添加。

（6）了解食品本身的物理、化学性质、加工、包装情况、贮藏条件及它们对防腐剂效果的影响，确定防腐剂的投放时机。

LESSON 5

FOOD RISK ANALYSIS AND FOOD QUALITY

1 FOOD RISK ANALYSIS

Risk analysis is a structured process. It is a systematic science. The Codex Alimentarius Commission (CAC) divides it into three parts: risk assessment, risk management, and risk communication; the three are repeated, continuous, and reciprocal feedback. Risk analysis is defined as that under the premise of reference related factors, describing the characteristics of the risk qualitatively or quantitatively, proposing and implementing risk management measures and exchanging relevant information by assessing various biological, physical and chemical hazards which affect food safety and quality. It is the basis for formulating food safety standards. Furthermore, hazards and risks are the two most basic concepts in food risk analysis, and it is important to clarify its definition for follow-up work. The CAC has a clear definition of the two. Hazards are the biological, chemical or physical factors or conditions in the food which may cause adverse health effects, while risks refer to the potential for adverse health effects and the severity of their effects, which leads to hazards in food.

2 RISK ASSESSMENT

Risk assessment is a scientifically based process consisting of the following steps:
Hazard identification;
Hazard characterization;
Exposure assessment;
Risk characterization.

In the framework of risk analysis, risk assessment is its scientific core. Risk assessment is, on the basis of existing scientific data, to identify, confirm, and quantify the adverse consequences of exposure to hazardous substances in food. It is the link to connect scientific knowledge and management measures, and the results will be applied to the formulation of relevant policies for food safety.

2.1 Hazard Identification

The identification of biological, chemical, and physical agents capable of causing adverse health effects and which may be present in a particular food or group of foods.

2.2 Hazard Characterization

Hazard characterization is the qualitative and/or quantitative evaluation of the nature of the adverse health effects associated with biological, chemical and physical agents which may be present in food. For chemical agents, a dose-response assessment should be performed. Dose-response assessment refers to the relationship between the degree of exposure (dose) of a chemical, biological or physical factor and the severity and/or frequency of related adverse health effects (responses). For biological or physical agents, a dose-response assessment should be performed if the data are obtainable.

2.3 Exposure Assessment

Exposure assessment is the qualitative and/or quantitative evaluation of the likelihood of biological, chemical, and physical intaking in food, as well as exposure assessment from other sources if relevant.

2.4 Risk Characterization

Based on hazard identification, hazard characterization and exposure assessment, risk characterization aims to evaluate the probability and severity of known or potentially adverse health effects in a given population qualitatively and/or quantitatively assessed, including concomitant uncertainties.

3 RISK MANAGEMENT

Risk management refers to the process of measuring policy choice based on risk assessment results. If necessary, select and implement appropriate control options, including regulatory measures.

On the basis of risk assessment, all stakeholders negotiate on various alternative food safety regulatory measures or programs, weigh the advantages and disadvantages, and ultimately select the most appropriate prevention or control program. The purpose is to control or reduce food safety risks as much as possible and safeguard public health. Risk management is mainly divided into four parts: ① Preliminary risk management activities, including identifying food safety issues and clarifying their nature, describing risk conditions, determining risk management objectives, determining whether to conduct risk assessments and developing assessment policies, conducting risk assessments, and conducting analysis and classification of results; ② Determination of risk management options, including determining alternative management measures, evaluating alternative management measures and selecting optimal management measures; ③ Implementing management measures, including verifying the effectiveness of necessary control systems and implementing control over selection measures and verification of the implementation; ④ Monitoring and assessing. The risk management process is the full exchange and coordination between risk managers, risk assessors and all stakeholders.

Risk Communication

Risk communication refers to the interactive exchange of information and opinions on risk-related content among risk assessors, risk managers, consumers, and other stakeholders throughout the risk analysis process.

The main purposes are: ① Promote active participation of stakeholders in the risk analysis process, and strengthen their understanding of the assessment issues, helping to improve the overall efficiency of the risk analysis process; ② Improve the consistency and transparency of risk management decision-making; ③ Promote stakeholders' understanding of risk management decisions to ensure smooth implementation of risk management decisions. Risk communication is a two-way process and is an indispensable part of the risk analysis framework.

4 FOOD QUALITY

After certain processing, food has a certain nutritional value and is harmless to the human body. Since the value of food is mainly reflected in its practicality, the quality of food can be defined as the characteristics that meet the customers' needs in terms of food edibility.

Quality Characteristics of Food

Food has quality characteristics. As tangible products, characteristics of food quality include functionality, credibility, safety, adaptability, economy and timeliness. As special products, characteristics of food quality are mainly reflected in the following three aspects:

Safety in the tangible product quality characteristics of food

Safety is an attribute that always needs to be considered first in the quality characteristics of foods. If the safety of food products is not enough, they will still lose their value as products and commodities even if other quality characteristics are good. The safety of food should ensure that the food does not contain toxic and harmful chemical substances or organisms (bacteria, viruses, parasites, etc.) that may damage or threaten human health, so as to avoid the risk of causing food-borne diseases to consumers.

Food products are functional and practical

In addition to the intrinsic performance and external performance, the functionality of food has potential cultural properties. Intrinsic properties include nutritional properties, flavor preference properties, and physiological regulation properties. External performance includes food modeling, style, color, luster and so on. Cultural performance includes ethnic, religious, cultural, historical, and customary characteristics (such as muslim food).

The comprehensive quality of food

In addition to the quality of tangible products, food quality includes process quality, service quality, and work quality. In food production, raw materials, production methods, production environment and other factors have a great influence on the quality of food. In addition, because food is a consumable product, its service quality is not reflected in the after-sales service, but reflected in the convenience of consumers to buy and use.

If there is a problem in the quality of the product, users will unavoidably have economic losses. However, if there is a problem with the quality of the food, then in addition to the economic loss, it may also be life-threatening for the user. With the development of society and economy and the acceleration of the pace of life, the proportion of people buying ready-to-eat foods will gradually increase. Therefore, the assurance and improvement of food quality is becoming increasingly important.

🕪 Lesson 5

Vocabulary

/ Words /

premise [ˈpremɪs] n. 前提
qualitatively [ˈkwɔliteitivli] adv. 定性地
quantitatively [ˈkwɔntɪtətɪvli] adv. 定量地
propose [prəˈpoʊz] v. 提出
relevant [reləvənt] adj. 相关的
framework [ˈfreɪmwɜːrk] n. 框架
adverse [ˈədvɜːrs] adj. 不利的
formulation [fɔːmjʊˈleɪʃn] n. 制定；规定
concomitant [kənˈkɑːmɪtənt] n. 伴随物
implement [ˈɪmplɪmənt] v. 实施
option [ɑːpʃn] n. 选项
intrinsic [ɪnˈtrɪnzɪk] adj. 本质的
external [ɪkˈstɜːrnl] adj. 外部的
luster [ˈlʌstər] n. 光泽
acceleration [əkˌseləˈreɪʃn] n. 促进

stakeholder [ˈsteɪkhoʊldər] n. 利益相关者
negotiate [nɪˈgoʊʃIeɪt] v. 商议
ultimately [ˈʌltɪmətli] adv. 最终
safeguard [ˈseɪfgɑːrd] n. 保护；护卫
preliminary [prɪˈlɪmɪneri] adj. 最初的
optimal [ˈɑːptɪməl] adj. 最理想的
verification [ˌverəfəˈkeɪʃn] n. 确认
monitor [ˈmɑːnɪtər] v. 监控
consistency [kənˈsɪstənsi] n. 一致性
transparency [trænˈspærənsi] n. 透明
indispensable [ˌɪndɪˈspensəbl] adj. 不可缺少的
organism [ˈɔːrgənɪzəm] n. 生物体
virus [ˈvaɪrəs] n. 病毒
parasite [ˈpærəsaɪt] n. 寄生虫
consumptive [kənˈsʌmptɪv] adj. 消耗性的

/ Phrases /

reciprocal feedback 相互反馈
hazard identification 危害识别
hazard characterization 危害特征描述
exposure assessment 暴露评估
risk characterization 风险特征描述
muslim food 清真食品

food-borne 食源性　　　　　　　　ready-to-eatfood 方便食品，即食食品
follow-up 后续的　　　　　　　　　after-sales 售后服务
dose-response 剂量反应　　　　　　two-way process 双向过程
life-threatening 威胁生命的　　　　　risk-related 风险相关的

Exercises

I. Write true or false for the following statements according to the passage.

1. (　) Risk analysis is a structured process. It is a systematic science and can be divided into three parts: risk assessment, risk management, and risk communication.

2. (　) Hazards and assessments are the two most basic concepts in food risk.

3. (　) In the framework of risk analysis, risk assessment is its scientific core.

4. (　) Qualitative and/or quantitative assessment of the likelihood of uptake of biological, chemical and physical factors by food is an exposure assessment.

5. (　) Risk management refers to the interactive exchange of information and opinions on risk-related content among risk assessors, risk managers, consumers and other stakeholders throughout the risk analysis process.

6. (　) The quality of food can be defined as the characteristics that the food meets the user's needs in terms of food quality.

7. (　) Food characteristics include packaging attributes.

8. (　) In addition to the quality of tangible products, food quality includes process quality, service quality, and work quality.

9. (　) The value of food is mainly reflected in the functional.

10. (　) The intrinsic properties of foods include nutritional properties, flavor preference properties, and physiological regulation properties.

II. Answer the following questions according to the passage.

1. What is a risk assessment?
2. What is risk analysis?
3. What is an exposure assessment?
4. What are the components of risk management?
5. What are the characteristics of food quality?

III. Fill in the blanks according to the passage.

1. Risk analysis is a structured process. It is a systematic science and is divided into three parts: _____, _____, and _____.

2. The _____ and _____ are the two most basic concepts in food risk. It is of great significance to define its definition for the follow-up work.

3. Risk assessment is a science-based process consisting of the following steps: _____,

_____, _____ and _____.

4. Food as a tangible product, its quality characteristics also include _____, _____, _____, _____, economy and timeliness.

5. The safety of food should ensure that the food does not contain _____ or _____ that may damage or threaten human health, so as to avoid the risk of causing _____ to consumers.

IV. Translate the following words and expressions into Chinese.

risk assessment   foodborne diseases
hazard identification   physiological regulation
hazard characterization   flavor preference performance
quantitative assessment   consumable products
risk communication   Codex Alimentarius Commission

V. Translate the following expressions into English.

1. 危害和分析是食品风险中最基本的两个概念，明确其定义对后续工作的进行有着重要意义。CAC 对这两者有明确的定义，危害是指食品中可能引起不良健康影响的生物、化学或物理因素或状况。风险是不良健康影响的可能性及其影响的严重程度，由此导致食品中的危害。

2. 食品质量除了有形产品的质量之外，还包括过程质量、服务质量、工作质量等内容。在食品生产中，原材料、生产方法、生产环境等因素对食品质量有很大的影响。此外，由于食品是消耗性产品，其服务质量不体现在售后服务方面，而体现在消费者购买和使用的方便性上。

参考译文

第 5 课　食品风险分析与食品质量

1　食品风险分析

风险分析是一个结构化的过程，是一门系统科学。国际食品法典委员会（Codex Alimentarius Commission，CAC）将其分为风险评估、风险管理和风险交流三个部分，三者循环往复、持续进行、相互反馈。风险分析的定义是：通过对影响食品安全质量的各种生物、物理和化学危害进行评估，定性或定量地描述风险的特征，在参考有关因素的前提下，提出和实施风险管理措施，并对有关情况进行交流，它是制定食品安全标准的基础。此外，危害和分析是食品风险中最基本的两个概念，明确其定义对后续工作的进行有着重要意义。CAC 对这两者有明确的定义，危害是指食品中可能引起不良健康影响的生物、化学或物理因素或状况。风险是不良健康影响的可能性及其影响的严重程度，由此导致食品中的危害。

2　风险评估

风险评估是一个基于科学的过程，由以下步骤组成：危害识别、危害特征描述、暴露评估和风险特征描述。

在风险分析框架中，风险评估是其科学核心。风险评估是利用现有的科学资料，对食品中的危害物暴露对人体健康产生的不良后果进行识别、确认和定量，是连接科学知识与管理措施的纽带，其结果将应用于食品安全的相关政策制定。

2.1　危害识别

识别能够对健康产生不利影响并可能存在于特定食物或食物组中的生物、化学和物理因素。

2.2　危害特征描述

危害特征描述是定性和（或）定量评估与食品中可能存在的生物、化学和物理因素有关的不利健康影响的性质。对于化学制剂，应进行剂量-反应评估。剂量-反应评估是指确定化学、生物或物理因素的接触程度（剂量）与相关的不良健康影响（反应）的严重程度和/或频率之间的关系。对于生物或物理因素，如果可以获得数据，应该进行剂量-反应评估。

2.3　暴露评估

暴露评估是对食物中生物、化学和物理因素摄取的可能性进行定性和（或）定量评估，以及如果相关的话，从其他来源的暴露评估。

2.4　风险特征描述

基于危害识别、危害特征描述和暴露评估，定性和（或）定量评估特定人群中已知或潜在不利健康影响的发生概率和严重程度，包括伴随的不确定性。

3　风险管理

风险管理是指根据风险评估结果衡量政策选择的过程，如果需要，选择和实施适当的控制选项，包括监管措施。

在风险评估的基础上，各利益相关方通过对各种备选的食品安全监管措施或方案进行磋商，权衡利弊，最终选择最适宜的预防或控制方案。目的在于尽可能控制或降低食品安全风险，保障公众健康。风险管理主要分为四个部分：①初步风险管理活动：包括识别食品安全问题并明确其性质、描述风险情况、确定风险管理目标、确定是否进行风险评估并制定评估政策、开展风险评估、进行结果分析及分级；②风险管理方案的确定：内容包括确定备选管理措施、评估备选管理措施和选择最优管理措施；③实施管理措施：包括验证必要控制体系的有效性、实施选择的控制措施并验证实施情况；④监控及评估：风险管理过程就是风险管理者、风险评估者及各利益相关者之间的充分交流与相互协调。

风险交流

风险交流是指在风险分析全过程中，风险评估人员、风险管理人员、消费者和其他利益相关方之间就风险相关内容进行信息和意见的交互式交流。

其目的主要有：①促进各利益相关方积极参与风险分析过程，加强各方对评估问题的理解，有利于风险分析过程中整体效率的提高；②提高风险管理决策制定的一致性和透明度；③促进各方对风险管理决策的理解，以保障风险管理决策的顺利实施。风险交流是一个双向的过程，是风险分析框架中必不可少却容易被忽略的部分。

4 食品质量

食品经过一定加工制作、具有一定营养价值且对人体无害。由于食品的食用价值主要体现在实用性上，因此，食品质量可定义为食品在食用性方面满足顾客需要的特性。

食品的质量特性

食品作为有形产品，其质量特性也包括功能性、可信性、安全性、适应性、经济性和时间性等，而作为特殊的产品，食品的质量特性，主要体现在以下三个方面。

食品有形产品质量特性中的安全性

安全性是食品的质量特性中需要首要考虑的属性，食品产品的安全性如果不过关，那么即使其他质量特性再好，也丧失了作为产品和商品存在的价值。食品的安全性应保证食品不含有可能损害或威胁人体健康的有毒有害化学物质或生物（细菌、病毒、寄生虫等），避免导致消费者患食源性疾病的危险。

食品的产品功能性和实用性

食品的功能性除了内在性能、外在性能以外，还有潜在的文化性能。内在性能包括营养性能、风味嗜好性能和生理调节性能。外在性能包括食品的造型、款式、色彩、光泽等。文化性能包括民族、宗教、文化、历史、习俗等特性（如清真食品）。

食品的综合质量

食品质量除了有形产品的质量之外，还包括过程质量、服务质量、工作质量等内容。在食品生产中，原材料、生产方法、生产环境等因素对食品质量有很大的影响。此外，由于食品是消耗性产品，其服务质量不体现在售后服务方面，而体现在消费者购买和使用的方便性上。

如果产品质量有问题，给消费者带来经济损失是不可避免的，但如果食品质量有问题，那么消费者除了经济损失之外，还可能会造成生命危险。随着社会经济的发展、生活节奏的加快，人们购买即食食品的比例会逐渐增加，因此食品质量的保证和提高显得越来越重要。

LESSON 6

FOOD SAFETY TRACING

In recent years, the frequent incidents of food safety problems in some countries have not only threatened the health of consumers in other countries and even the world, but also weakened their confidence in food safety and the integrity of certain food, even in national and regional management systems due to the integration of agricultural production and food industry as well as the globalization of food trade. Therefore, it is imperative to strengthen the management of food quality and safety, ensure food quality and safety, enhance consumers' trust, establish a food traceability system and realize the traceability supervision of food.

1 FOOD GEOGRAPHICAL ORIGIN TRACING

Food origin tracing is the process of identifying a source of a particular product or a batch of products from the downstream of supply chain to the upstream. Food geographical origin tracing means building the guidelines, techniques, means and documents required by food companies for establishing food trace, as well as by the supervision department in tracing, confirming and recalling. Food authenticity is the consistency between the food purchased by the consumer and the trademark or product description, which proves the authenticity of the product. Food geographical origin authenticity is to verify whether the food actually comes from a certain place in an objective way. It is more difficult to determine which region the food comes from than whether the food is from a certain region. That is to say, it is a difficulty to realize food geographical origin tracing than the food geographical origin authenticity. Food geographical origin tracing can be achieved by tracing information such as paper, pen, electronic label and ear tag. In case of information being lost or forged, it can be traced through stable isotope fingerprint, mineral element fingerprint, infrared spectrum fingerprint, organic component fingerprint analysis and so on.

1.1 The Development Process of Food Geographical Origin Tracing

Food geographical origin tracing originated in France, used to ensure the quality of wine and champagne when there were no laws and regulations to enforce it. In the early 21st century, Europe's "mad cow disease" made food tracing become increasingly important. Hence in 2002, European Union rule (No. 178/2002) claimed that all foods sold within the European Union should be tracked and traced since 2005. In 2006, the European Community ordinance (No. 510/2006) required that protection of the geographical and appellate names of agricultural products and food should be implemented. In 2008, Administrative Measures on the Geographical Indication of Agricultural Products was carried out. In 2009, Food Safety Enhancement Act of 2009 required that all non-processed foods should be tagged with labels to identify the origin in the USA. In China, the food safety supervision system has been built according to Food Safety Law of PRC since 2009, and then this law was revised and referred to the regulatory requirements related to traceability repeatedly in 2015. In the same year, it was required to improve the quality of agricultural products and the level of food safety, as well as establish a platform for data information sharing through the Internet according to the Central Document No. 1.

1.2 Fingerprint Technology of Food Geographical Origin Tracing

Food fingerprinting technology refers to the analysis of the commonness and personality of various components in food through proper treatment of the samples, and the different characteristics of different components can be used to express different characteristics (including spectrum, chromatography, mass spectrometry and other spectra), and then the chemical composition and properties of food are well represented by measuring the characteristics of food. In combination with stoichiometry, the evaluation of food is more objective and comprehensive, which is also integrated, hierarchical, associative, etc. It not only depends on the advanced analysis techniques, but also depends on the mathematical statistics method and computer simulation methodology. The food quality fingerprint based on a lot of fundamental work is an effective reference to identify food security. Fingerprint, as a new quality control mode, is widely regarded, and the application of this technology will play an important role in the quality control of food and the inspection of food quality.

2 CLASSIFICATION OF FOOD GEOGRAPHICAL ORIGIN TRACING

2.1 Main Techniques of Fingerprint Tracing

It is very important for the domestic development of fast and accurate tracing technology of agricultural products due to the counterfeiting and shoddy of excellent agricultural products as well as the protection of original products. Agricultural origin traceability is mainly analyzing characterization to identify different specific indicators of regional source of agricultural products. Nowadays, the fingerprint map to distinguish the origin of agricultural products is established based on techniques including the mineral elements fingerprint analysis technology, electronic nose fingerprint technology, near infrared spectral analysis technology, DNA fingerprint techniques, metabonomics fingerprint technology combined with stoichiometry study, so it can trace back to the origin of different kinds of agricultural products by establishing a characteristic fingerprint that can distinguish the origin of agricultural products.

2.2 Traceability Technology of Mineral Elements

Organisms need to intake minerals from the surrounding environment, which cannot be synthesized in their bodies and are effected by local water, geological factors, soil environment, etc. Thus, there are distinctive fingerprint characteristics of mineral elements in the organisms of different regions. This technology can select the efficient features by analyzing the composition and content of the different source of mineral elements in organisms with mathematical statistics methods like variance analysis, clustering analysis and discriminant analysis. Then the discriminant model and database are established to realize food traceability and confirmation. Haiyan Zhao et al. showed that the fingerprint analysis technology of mineral elements can be used in the discrimination of wheat producing area. Jie Wang et al. proved that mineral elements, as a quality index of tea, were related to its quality and safety; at the same time the tea mineral elements carrying the regional characteristic fingerprint information can be used as an important landmark used in origin tracing.

2.3 Electronic Nose Fingerprint Technology

The electronic nose is also known as the scent scanner, which can quickly determine the overall information of volatile components in samples through a specific sensor array, signal processing and pattern recognition system. Guixian Hu et al. used the electronic nose system, equipped with 10 sensor arrays to analyze maturity of 5 batches of citrus picked at different times (15d/batch) and the correct judgment rate was 88%. This provides a basis for judging citrus maturity. Compared with traditional chemical analysis methods and sensory evaluation methods, electronic nose fingerprint technology has the advantages of simple operation, simple pretreatment, quick determination and good reproducibility. However, there are still many problems that need to be further studied in terms of manufacturing materials, testing environment and data processing.

2.4 Near-infrared Spectrum Traceability Analysis Technique

Near infrared and infrared are kinds of rapid and nondestructive testing techniques, based on the non-stop vibration and rotating of molecular. Each molecular has its unique infrared absorption spectrum determined by its composition and structure, thus it can be used to analyze and authenticate molecular structure. Ning Zhang et al. established a traceability model for mutton in Shandong, Hebei, Inner Mongolia and Ningxia, using near infrared spectrum in SIMCA pattern recognition method after pretreated with 5 point smoothing and multiple scattering, and the recognition rate of the model were 100%, 83%, 100% and 92% respectively. Ping Zhang et al. studied the adulteration of edible oil using near infrared spectroscopy, which distinguished three kinds of cooking oil and sesame oil mixed with other fats clearly. Quansheng Chen et al. established models of tea validation of Longjing, Biluochun, Qihong and Tieguanyin and used near-infrared spectroscopy combining with SIMCA pattern recognition method, and the recognition rates were 100%, 83%, 100% and 92%, respectively.

2.5 DNA Fingerprinting Technology

DNA fingerprint is a kind of genetic marker based on the variation of nucleotide sequence among organisms that can directly detect the differences between individual organisms at DNA level, and it is a direct reflection of genetic variation at DNA level. The fingerprint is a powerful tool to identify varieties and strains (including hybrid parent and inbred lines) with the advantages of rapidity, accuracy, etc. Its main purpose is variety improvement and seed identification, and it is very suitable for identification of food adulteration and quality tracing of variety resources. Some of the existing studies have played an increasingly important role in the identification of fish, alfalfa, rice, corn, wheat, meat products, fruit juices, etc.

2.6 Metabolomics Fingerprint Technique

Metabolomics fingerprint technique is a kind of comprehensive qualitative and quantitative analysis method conducted without explicit target compounds, which can be used to detect the whole metabolites of a certain bio-sample under certain chromatographic conditions. The aim is to quantitatively describe the whole endogenous metabolites, and the response rules of internal and external changes. The main method is the qualitative and quantitative analysis, the main metabolite target analysis, the metabolic profile analysis, metabolic fingerprint analysis and metabolomics analysis to all small molecule metabolites of the whole cells in a biological system, within a given time and condition.

3 THE APPLICATION OF FINGERPRINT TECHNOLOGY IN THE TRACING OF AGRICULTURAL PRODUCTS

The source tracing technology of agricultural products is one of the important technical supports for the protection of geographical indication products, which can provide the theoretical basis for the tracing and confirmation of geographical indication products and regional famous products. The effective method is to select the characteristic factor which can represent the regional information, and analyze its "fingerprint" characteristics through stoichiometry, so as to identify the origin of food. Baoxin Lu determined the effective index of the origin of soybean in Heilongjiang using the mineral element fingerprint technique, and As, Ru, Gd and Tb, as reliable indicators were selected from 46 kinds of mineral elements. Pinghui Li et al. used the fingerprint technique of mineral elements to determine the origin of kidney bean, and selected Ca, As, Mg and Pt 4 elements to establish the origin tracing model of kidney bean. Yanan Zhao et al. carried out genetic diversity analysis and screening to 14 pairs of SSR primers for 35 green bean varieties in Heilongjiang province and constructed green bean SSR fingerprints and molecular identity cards, using the SSR fluorescence labeling technique. This achievement can be used for the identification of green bean varieties.

3.1 Research on Fingerprint Tracing Technology of Japonica Rice

It is very important to study the fingerprint information technology of rice origin and establish the rice fingerprint database for the soundness and development of the food security control system and the corroboration system in Heilongjiang province. Due to the unique quality and influence of geographical indications of brand agricultural products (GI products) under the drive of economic interests, the prices of these products are much higher than similar products in the market. To promote the management of GI products, it is imperative to study thoroughly the influencing factors of fingerprint information which is unique to the region of GI products, establishing discriminant models and constructing databases to enhance the enforceability of relevant laws and regulations, and to promote the application of tracing techniques in practice.

3.1.1 Feasibility study of the tracing technology of mineral elements in rice

The research on the origin of domestic food sources mainly focuses on the preliminary stage of the feasibility of mineral element fingerprints on food origin tracing. The contents of mineral elements in different parts of rice can be found significantly different, which resulted in the loss of K, Ca, Na and other elements after milling through using X-ray photoelectron spectrometer to test the rice. At the same time, the content and composition of mineral elements in agricultural products carry different information of its origin. So it is feasible to identify the origin of rice by means of mineral element fingerprint analysis.

(1) Variation of mineral element content in japonica rice by variety, region and processing precision

The content of mineral elements in rice is affected by many factors. The result that rice variety, region and processing precision can influence the content of mineral elements in japonica rice provides a theoretical reference for the application of fingerprint tracing technology of mineral elements in agricultural products.

(2) Analysis of soil mineral elements in different regions

The content of mineral elements in food is closely related to its habitat. That is to say, the distribution of mineral elements in soil, water source and other resources are different, which results in a more stable and effective reflection into agricultural products. The correlational analysis of mineral elements in the soil of food and origin will make the tracing method more reliable. The regions of rice can be distinguished significantly through analyzing the element where rice is closely related to maternal soil.

(3) Discriminant analysis of mineral element contents closely related to region in different regional japonica samples

The mineral elements needed for rice are mainly absorbed from the soil. The distribution of trace elements in soil affects the quality of rice. The information carried by closely related elements can distinguish the origin of rice. The discriminant model was established using closely related elements, and the correct classification rate was higher.

3.1.2 Construction of the Northern Japonica Fingerprint Map Based on AFLP

The northern DNA fingerprint database can be constructed by establishing a molecular marker system of amplified fragment length polymorphism (AFLP), which is applicable to the northern japonica varieties. This technology, combined the advantages of RFLP and RAPD technology, is becoming both reliable and convenient. Its principle is that a random restricted DNA fragment with different molecular weight is formed cut by plant genomic DNA enzyme, and then the joint of the specific coenzyme point is connected to the two ends of the enzyme cutting section, after that PCR amplification, electrophoresis, radioactivity or non-radioactivity, DNA fingerprinting were performed using the primers of the specific complementary joint. We can reveal the affinity of different varieties of japonica rice, and can distinguish the japonica rice varieties with different relatives using AFLP molecular marker method.

3.2　Research on Fingerprint Tracing of Miscellaneous Grains

DNA fingerprints tracing technology is mainly aimed at containing genes of genetic material, analyzing grains' base sequence differences in gene, finally achieving the purpose of identification, and distinction between grain varieties.

3.2.1　Diversity Analysis

Diversity mainly refers to the sum of genetic variation between different groups of different living groups or individuals. It provides an important basis for the full utilization, scientific breeding and genetic research of miscellaneous grain resources to analyze the genetic diversity of germplasm resources and understand the genetic distance between populations. In particular, the variety of millet varieties and the establishment of a database of origin provide the possibility for tracing and protecting of origin of excellent millet.

3.2.2　Establishment of Fingerprint Map

The genetic map is a linear arrangement of genes obtained by genetic recombination, which illustrates the relative relationship between a few functional genes and genetic markers. SSR markers are generally used as anchors because of its co-dominant, locus stable and polymorphic high characteristics that are beneficial to integrate different chain groups and map linkage groups or chromosomes. Genetic linkage map based on SSR marker technique is helpful to further understand the agronomic characters of grains at the molecular level, which has great significance to study gene localization, cloning, genome structure and function as well as genetic analysis of important traits.

3.2.3　Genetic Map Construction

The genetic map refers to the chromosomal linear linkage map with the chromosome recombination exchange rate as the relative length unit and the genetic marker as the main body. The construction of genetic maps is an important link in genome research, also the basis of gene mapping and cloning even the structure and function of the genome. The genetic map of legumes DNA has always been a hot topic in the world.

3.2.4　Species Identification

Along with molecular biology, especially transgenic technology widely used in the field of crop breeding, there are a lot of similarities and genetic differences in morphology of new varieties, which makes it impossible for the traditional morphological methods to be accurately identified. At the same time, the domestic seed market part of the phenomenon of poor quality deepened the difficulty of variety identification. From the perspective of genetics, variety purity and authenticity identification are essentially the identification of genotypes. Therefore, the identification of plant varieties is more accurate and reliable by the DNA labeling technology of the DNA molecules directly identified.

4 RESEARCH PROSPECT OF BIOLOGICAL FINGERPRINT TRACING TECHNOLOGY

At present, origin of agricultural products is closely related to its quality, safety and nutritional quality. Origin information is the important basis for consumers to choose agricultural products, and it is also an effective way to ensure food safety where the farmers and agricultural enterprises pay much attention. The construction of China's agricultural product quality traceability system started late. Besides, due to small agricultural production unit scale, large quantities, low level of informationization and other national conditions, the tracing development is also still very slow. However, with the rapid improvement of the life quality of consumers, the demand for agricultural products and attention have changed from quantity to quality. In recent years, China has entered the rapid development period of agricultural product quality and safety tracing. Various biological fingerprint techniques with their high sensitivity, fast analysis speed, and low detection limit, spring up in aspects of the effective identification of origin of agricultural products, ensuring the quality and safety of agricultural products and maintaining the agricultural enterprises' good image. It not only guarantees the consumer's right to know, but also promotes the benign development of agriculture.

However, there are still many problems to be solved and improved. For example: how to screen the source tracing indicators effectively; the number of sample collection and the determination of the range of tracing; a variety of origin-tracing technology fusion research; construction of an effective food source tracing data sharing platform; further development of corroboration technology; the cost of tracing technology and so on. With the unremitting work of agricultural researchers and the rapid development of agriculture in China, the technology will become mature and perfect and will be widely used and developed in all kinds of agricultural products and food.

◁⸺)) Lesson 6

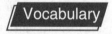

/ Words /

traceability [ˌtreɪsəˈbɪlɪti] n. 溯源
fingerprint [ˈfɪŋɡərprɪnt] n. 指纹
management [ˈmænɪdʒmənt] n. 监管
spectrum [ˈspektrəm] n. 光谱

chromatography [ˌkroʊməˈtɑːgrəfi] n. 色谱
characteristic [ˌkærəktəˈrɪstɪk] n. 特征
stoichiometry [ˌstɔɪkɪˈɒmətrɪ] n. 化学计量学
simulation [ˌsɪmjuˈleɪʃn] n. 模拟
synthesize [ˈsɪnθəsaɪz] v. 综合
nondestructive [ˌnɒndɪˈstrʌktɪv] adj. 无损的

explicit [ɪkˈsplɪsɪt] adj. 明确的
metabolite [meˈtæbəlaɪt] 代谢产物
enforceability [ɪnˌfɔːsəˈbɪlətɪ] n. 可执行性
feasibility [ˌfizəˈbɪləti] n. 可行性
nucleotide [ˈnukliətaɪd] n. 核苷酸
morphology [mɔːrˈfɑːlədʒi] n. 形态学
genotype [ˈdʒenətaɪp] n. 基因型

/ Phrases /

mathematical statistics method 数理统计方法
effective reference 有效参考
fingerprint trace 指纹溯源
food geographical origin tracing 食品产地溯源技术
food authenticity 食品确证
food geographical origin authenticity 食品产地确证
the mineral elements fingerprints analysis technology 矿物元素指纹分析技术
electronic nose fingerprint technology 电子鼻指纹图谱技术
near infrared spectral analysis technology 近红外光谱溯源分析技术
DNA fingerprint techniques DNA 指纹图谱技术
metabonomics fingerprint technology 代谢组学指纹图谱技术
variance analysis 方差分析
clustering analysis 聚类分析
discriminant analysis 判别分析
volatile components 挥发性成分
sensor array 传感器阵列
pattern recognition system 模式识别系统
infrared absorption spectrum 红外吸收光谱
mass spectrometry 质谱
X-ray photoelectron spectrometer X 射线光电子能谱仪
diversity analysis 多样性分析
genetic map 遗传图谱

/ Abbreviations /

deoxyribonucleic acid (DNA) 脱氧核糖核酸
amplified fragment length polymorphism (AFLP) 扩增片段长度多态性
restriction fragment length polymorphism (RFLP) 限制性内切酶片段长度多态性

random amplified polymorphic DNA (RAPD) 随机扩增多态性 DNA 标记
polymerase chain reaction (PCR) 聚合酶链式反应
simple sequence repeat (SSR) 简单重复序列标记

Exercises

Ⅰ. Write true or false for the following statements according to the passage.

1. (   ) Food trace is the process of identifying a particular product or product source downstream from the supply chain.

2. (   ) The electronic nose is also known as the scent scanner, which can quickly determine the overall information of volatile components in samples through a specific sensor array, signal processing and pattern recognition system.

3. (   ) In China, food safety supervision system should be built according to Food Safety Law of PRC since 2008.

4. (   ) In combination with stoichiometry, the differences and quality characteristics of food are evaluated more objectively and comprehensively, the characteristics of comprehensiveness, integrity, hierarchy and relevance.

5. (   ) The content of mineral elements in rice is affected by a few factors.

6. (   ) Near infrared and infrared are kinds of rapid and nondestructive testing techniques, based on the non-stop vibration and rotating of molecular, which has its unique infrared absorption spectrum determined by its composition and structure, thus it can be used to analyze and authenticate molecular structure.

7. (   ) We cannot reveal the affinity of different varieties of japonica rice, and cannot distinguish the japonica rice varieties with different relatives using AFLP molecular marker method.

8. (   ) The genetic map is a linear arrangement of genes obtained by genetic recombination, which illustrates the relative relationship between many functional genes and genetic markers.

9. (   ) The source traceability technology of agricultural products is one of the important technical support for the protection of geographical indication products, which can provide the theoretical basis for the traceability and confirmation of geographical indication products and regional famous products.

10. (   ) AFLP combines the advantages of RFLP and RAPD technology, becoming both reliable and convenient.

Ⅱ. Answer the following questions according to the passage.

1. What is the food geographical origin tracing?

2. What do main techniques of fingerprint trace mainly include?

3. When did European Union rule (No. 178/2002) claimed that all foods sold within the European Union could be tracked and tracked?

4. What is DNA fingerprint techniques?

5. What is the problem to be solved and improved of biological fingerprint tracing technology?

Ⅲ. Fill in the blanks according to the passage.

1. In combination with stoichiometry, the differences and quality characteristics of food are evaluated more _____ and _____, the characteristics of _____, integrity, hierarchy and relevance.

2. Nowadays, a fingerprint map to distinguish the origin of agricultural products is going to be established based on the techniques including _____, electronic nose fingerprint technology, _____, DNA fingerprint techniques, _____ combined with stoichiometry study, so as to trace back to the origin of different kinds of agricultural products.

3. _____ is also known as the scent scanner, which can quickly determine the overall information of volatile components in samples through a specific sensor array, signal processing and pattern recognition system.

4. _____ is one of the important technical support for the protection of geographical indication products, which can provide the theoretical basis for the traceability and confirmation of geographical indication products and regional famous products.

5. _____ is a linear arrangement of genes obtained by genetic recombination, which illustrates the _____ between a few functional genes and genetic markers.

Ⅳ. Translate the following words and expressions into Chinese.

mathematical statistics method   volatile components
effective reference   diversity analysis
clustering analysis   sensor array
mass spectrometry   infrared absorption spectrum
electronic nose fingerprint technology

Ⅴ. Translate the following expressions into English.

1. 随着分子生物学，特别是转基因技术在农作物育种领域的广泛应用，大量在形态上相似而基因却存在差异的新品种出现，导致传统形态学方法不能准确鉴定品种。

2. DNA指纹溯源技术在杂粮品种鉴定中主要也是针对含有遗传物质的基因片段，分析杂粮基因中的碱基序列差异，最终达到鉴定、区分杂粮品种的目的。

参考译文

第6课　食品安全溯源

近年来，农业生产和食品工业的一体化以及食品贸易全球化趋势，导致如果一些国家或地区的重大食品安全事件频发和普发，将不仅直接威胁其他国家甚至全世界消费者的健康，更进一步削弱消费者对食品安全及某些食品的信任度，乃至国家及地区管理体系信心。因此，强化食品质量安全的管理，保障食品质量安全，提升消费者的信任，建立食品可追溯体系，实现食品的可追溯监管，已势在必行。

1 食品产地溯源技术体系

食品溯源是指从供应链下游向上游识别一个特定产品或一批产品来源的过程。食品产地溯源技术体系是指食品企业在建立食品追踪时所需要的指南、技术、手段和文件，以及监管部门在溯源、确证和召回时所需要的指南、技术、手段和文件。食品确证是指消费者购买的食品与其商标或产品说明的一致性，即证明产品的真实性。食品产地确证是通过客观的方法证实食品是否真正来自某个地区。明确食品来自哪个地域的难度要大于确证食品是否来自某个地域，即食品产地溯源技术的实现难度大于食品产地确证技术的实现难度。食品产地溯源可通过纸笔、电子标签、耳标等跟踪信息的方式实现。在跟踪信息丢失或伪造的情况下，可通过稳定同位素指纹、矿物元素指纹、红外光谱指纹和有机成分指纹分析等技术进行溯源。

1.1 食品产地溯源体系的发展进程

食品产地溯源技术起源于法国，最早用于保证葡萄酒和香槟酒的品质，但当时并没有相关法律法规对其进行强制要求。21世纪初欧洲"疯牛病"肆虐，使得食品可追溯的重要性日益显著。因此，2002年欧盟178/2002法规要求，从2005年起，在欧盟范围内销售的所有食品都能够进行跟踪与追溯；2006年欧共体第510/2006号条令要求对农产品和食品的地理标志和原产地名称实施保护；2008年实施了《农产品地理标志管理办法》，2009年美国《2009年食品安全加强法案》要求所有非加工食品须附有标签，标识原产地。我国2009年通过的《中华人民共和国食品安全法》中要求建立食品安全监管体制；2015年修订《中华人民共和国食品安全法》中多次提及与溯源有关的法规要求；2015年中央"一号文件"中明确指出提升农产品质量和食品安全水平，建立全程可追溯、互联网共享数据信息平台。

1.2 食品产地溯源体系的指纹图谱技术

食品指纹图谱技术是指样品经适当处理后，通过对食品中各个成分的共性和个性分析，依据不同的成分表达不同的特征谱学（包括光谱、色谱、质谱和其他谱等），通过测定食品的图谱特征，从而能够很好地表示其化学成分组成和特性。结合化学计量学方法，更加客观、全面地评价食品间的差异和品质特征，具有整体性、层次性、关联性等特点。其不仅依赖于先进的分析技术，而且还与数学统计手段以及计算机模拟的方法学密切相关，结合大量基础工作建立的食品质量指纹图谱是辨别食品防伪的有效参考。指纹图谱这一新的质量控制模式已受到广泛的重视，应用该项技术对食品的质量控制及食品的品质检测将会起到重要的作用。

2 食品产地溯源技术的分类

2.1 指纹溯源的主要技术

原产地保护产品和名优农产品以假乱真、以次充好的严重现象，使得国内发展快速准确的农产品产地溯源技术意义十分重大。农产品产地溯源主要是分析表征不同地域来源农产品的特异性指标，目前主要采用矿物元素指纹分析技术、电子鼻指纹图谱技术、近红外光谱溯源分析技术、DNA指纹图谱技术、代谢组学指纹图谱技术结合化学计量学研究，建立起能区分农产品产地来源的特征指纹图谱，从而对不同种类农产品进行产地溯源。

2.2 矿物元素产地溯源技术

生物体需要从周围环境中摄取自身不能合成的矿物元素，而矿物元素又受当地水、地质因素、土壤环境等的影响，导致不同地域生长的生物体内有各自的矿物元素指纹特征。该技术通过分析不同来源生物体中矿物元素的组成和含量，再利用方差分析、聚类分析和判别分析等数理统计方法筛选出有效指标，进而建立判别模型和数据库，实现食品溯源和确证。赵海燕等研究结果表明矿物元素指纹分析技术可用于小麦产地的判别。王洁等探明茶叶矿质元素作为茶叶品质指标关系着茶叶的品质与质量安全，同时茶叶矿质元素携带着地域特征指纹信息可以作为重要的标志性物质用在产地溯源中。

2.3 电子鼻指纹图谱技术

电子鼻又称气味扫描仪，是通过特定的传感器阵列、信号处理和模式识别系统，快速测定样品中挥发性成分的整体信息。胡桂仙等运用10个传感器阵列的电子鼻系统对5批不同时间（15天/批）采摘的柑橘进行成熟度检测分析，正确判断率达88%，为判断柑橘成熟度提供了依据。与传统的化学分析方法和感官评价方法相比，电子鼻指纹图谱技术，具有操作简便、样品前处理简单、测定速度快、重现性好等优点，但受制造材料、检测环境、数据处理等方面的限制，仍有许多问题需要进一步研究。

2.4 近红外光谱溯源分析技术

近红外和中红外属快速、无损检测技术，是由分子不停地振动和转动而产生的，而每种分子都有由其组成和结构所决定的独特的红外吸收光谱，因此可以对分子进行结构分析和鉴定。张宁等在830~2500nm波长范围内，近红外光谱经5点平滑与多元散射校正预处理，采用SIMCA模式识别方法对山东、河北、内蒙古、宁夏4个产地的羊肉建立产地溯源模型，模型对验证集样品的识别率分别为100%、83%、100%、92%。张萍等采用近红外光谱技术对食用油掺假问题进行了研究，可以清晰地区分3种食用油和掺杂了其他油脂的芝麻油。陈全胜等采用近红外光谱结合SIMCA模式识别方法，分别对龙井、碧螺春、祁红和铁观音4类茶叶验证的建立模型识别率分别是90%、80%、100%和100%。

2.5 DNA 指纹图谱技术

DNA 指纹图谱是指以生物个体间核苷酸序列变异为基础的遗传标记，可直接在 DNA 水平上检测生物个体间的差异，是生物个体在 DNA 水平上遗传变异的直接反映。指纹图谱是鉴别品种、品系（含杂交亲本、自交系）的有力工具，具有快速、准确等优点，多以品种改良和种子鉴定为目的，非常适合于品种资源的食品掺假和质量溯源的鉴定研究。现有的一些研究在鱼类、苜蓿、水稻、玉米、小麦、肉制品、果汁鉴定等方面，发挥着越来越重要的作用。

2.6 代谢组学指纹图谱技术

代谢组学指纹图谱技术，即在不明确目标化合物的前提下全面定性定量分析，一定色谱质谱条件下的某生物样品通过上述方法能够检测到其中的全部代谢产物。其目的是定量描述整个生物内源性代谢物质，及内因和外因变化的响应规则。而主要方法是对一个生物系统的全部细胞在给定时间和条件下所有小分子代谢物质的定性定量分析、主要代谢物靶标分析、代谢轮廓分析、代谢指纹分析和代谢组学分析。

3 指纹图谱技术在农产品溯源方面的应用

农产品产地溯源技术是保护地理标志产品的重要技术支撑之一，能够为地理标志产品、地区名特优产品的追溯和确证提供理论依据。有效方法是选择能够表征地域信息的特征因子，并通过化学计量学方法分析其"指纹"特征，从而识别食品的原产地。鹿保鑫利用矿物元素指纹图谱技术判别黑龙江黄豆产地溯源的有效指标，46 种矿物元素可筛选出 As, Ru, Gd, Tb 4 种作为可靠指标准确判别。李平惠等利用矿物元素指纹图谱技术判别芸豆产地来源，筛选出 Ca, As, Mg, Pt 4 种元素建立芸豆产地溯源模型。赵雅楠等利用 SSR 荧光标记技术，对黑龙江地区 35 份绿豆品种的 14 对 SSR 引物进行遗传多样性分析和筛选并构建了绿豆 SSR 指纹图谱及分子身份证，可用于绿豆品种真伪的鉴定。

3.1 地理标志水稻指纹溯源技术研究

研究水稻产地溯源指纹信息技术，建立水稻指纹数据库对黑龙江省粮食安全控制体系和确证体系的健全和发展至关重要。由于地理标志品牌农产品（GI 产品）在经济利益驱动下的独特品质和影响力，这些产品的价格远高于市场上的同类产品。为了促进地理信息产品的管理，必须深入研究地理信息产品区域特有的指纹信息的影响因素。建立判别模型，构建数据库，以增强相关法规条例的可执行性，促进溯源技术在实际中的应用。

3.1.1 水稻中矿物元素产地溯源技术的可行性研究

国内食品产地溯源研究主要集中在探讨矿物元素指纹对食品产地溯源的可行性的初级阶段，利用 X 射线光电子能谱仪测试水稻，会发现矿物元素在水稻不同部位的含量差异显著，碾磨加工后导致 K、Ca、Na 等元素丢失。同时，农产品中矿物质元素含量和组成携有产地信息，不同地域来源农产品中矿物元素含量和组成不同，通过矿物元素指纹分析技术鉴别水稻产地具有一定的可行性。

（1）品种、地域和加工精度对粳稻中矿物元素含量变异影响　稻米中的矿物元素含量受多种因素影响。稻米的品种、地域、加工精度会对粳稻中矿物元素含量产生影响的结论，为农产品产地矿物元素指纹溯源技术的应用提供理论参考。

（2）不同地域土壤矿物元素含量分析　食品中矿物元素的含量与其产地环境关系密切，不同产地土壤、水源等资源中矿物元素的分布存在差异，进而更加稳定和有效地反映到农产品中。通过食品与产地土壤中元素的相关性分析筛选元素组将使溯源方法更加可靠，运用稻米与母质土壤密切相关的元素进行判别分析，能显著区分各个地域。

（3）不同地域粳稻样品中与地域密切相关的矿物元素含量的判别分析　稻米所需要的矿物元素主要从土壤中吸取，土壤中微量元素的分布对稻米的品质产生影响的各种实验结论，力证与地域密切相关的元素携带的信息可以区分鉴别稻米的产地，采用与地域密切相关元素进行逐步判别分析，建立的判别模型，样品交叉验证正确分类率较高。

3.1.2　基于 AFLP 对北方粳稻指纹图谱的构建

通过建立适用于北方粳稻品种的扩增片段长度多态性（Amplified fragment length polymorphism，AFLP）分子标记体系可以构建北方 DNA 指纹数据库。该技术结合了 RFLP 和 RAPD 技术的优点，既可靠，又方便。其原理是植物基因组 DNA 酶切后，形成分子质量大小不等的随机限制性 DNA 片断，将特定同酶切点互补的接头连接在酶切片断的两端，然后用特异同接头互补的引物进行 PCR 扩增、电泳、放射性或非放射性显示 DNA 指纹。利用 AFLP 分子标记手段可以揭示不同品种粳稻的亲缘关系，能够区分开不同亲缘关系的粳稻品种。

3.2　杂粮指纹溯源技术研究

DNA 指纹溯源技术在杂粮品种鉴定中主要也是针对含有遗传物质的基因片段，分析杂粮基因中的碱基序列差异，最终达到鉴定、区分杂粮品种的目的。

3.2.1　多样性分析

多样性主要是指种内不同居群之间或同一居群不同个体之间的遗传变异的总和。对杂粮种质资源的遗传多样性进行分析，了解种群之间的遗传距离，为杂粮资源的充分利用、科学育种及遗传学研究提供重要依据。尤其是黍类品种多样性与原产地构建数据库，为黍类名优品种溯源及原产地保护提供了可能性。

3.2.2　指纹图谱构建

遗传图谱是由遗传重组所得到的基因线性排列图，说明了少数功能基因与遗传标记之间的相对关系。一般将 SSR 标记作为锚定标记，因其共显性、位点稳定且多态率高等特点，有益于整合不同连锁群和图谱连锁群或染色体的归并。基于 SSR 标记技术所构建的黍类遗传连锁图谱，更有利于从分子水平上对杂粮的农艺性状进行进一步了解，对基因定位、克隆、基因组结构与功能的研究以及重要性状的遗传分析具有重要的意义。

3.2.3　遗传图谱构建

遗传图谱是指以染色体重组交换率为相对长度单位，以遗传标记为主体的染色体线状连锁图谱。遗传图谱的构建是基因组研究中的重要环节，也是基因定位与克隆乃至基因组结构及功能研究的基础。豆类 DNA 遗传图谱的构建在国际上一直都是热点。

3.2.4 品种鉴定

随着分子生物学,特别是转基因技术在农作物育种领域的广泛应用,大量在形态上相似而基因却存在差异的新品种出现,导致传统形态学方法不能准确鉴定品种。同时,国内种子市场部分以劣充优的现象,加深了品种鉴定的难度。从遗传学角度上来看,品种纯度和真伪鉴定实质上是对品种基因型的鉴定。因此,通过直接鉴定品种的 DNA 分子本身的 DNA 标记技术则对植物品种的鉴定更为准确可靠。

4 生物指纹溯源技术研究展望

现阶段,农产品的产地来源与其质量、安全以及营养品质密切相关,农产品原产地信息是消费者选择购买农产品的重要依据,也是保证食品安全的有效途径,受到农民及农产品企业的密切关注。因我国农产品质量追溯体系制度建设起步较晚,农业生产单位规模小、数量大、信息化程度低等现实国情,使得追溯发展缓慢。但随着消费者生活品质的迅速提高,对农产品的需求和关注发生了由量到质的转变。近年来我国进入到了农产品质量安全追溯体系的快速发展时期,各种生物指纹图谱技术凭借其灵敏度高、分析速度快、检出限低等优点,在有效鉴定农产品的产地、保证农产品的质量与安全、维护农产品企业良好形象方面异军突起,既保证了消费者的消费知情权,更促进了农业的良性发展。

但目前对生物指纹溯源技术的研究,仍有许多亟待解决与改进的瓶颈。例如,如何有效筛选产地溯源指标的问题;样本采集数量及溯源研究范围的确定;多种产地溯源技术融合研究共进;建设有效的食品产地溯源数据共享平台;进一步发展确证技术的研究;溯源技术的成本问题等。相信随着广大农业科研工作者的不懈研究和我国农业的快速发展,该技术将日趋成熟与完善,在各类农产品和食品中得到广泛应用和发展。

LESSON 7

FOOD SAFETY AND MICROBIAL INFLUENCE FACTORS

1 MICROORGANISMS AND FOOD SAFETY

Microorganisms may cause food spoilage or food borne illnesses. Food spoilage means the undesirable changes in the food odor, color, taste, texture and appearance, or even the loss of its edible value. Some microorganisms do not directly cause changes in food but may alter the flora so that spoilage organisms can grow.

Food borne illnesses are usually caused by either microorganisms or toxins in food, and include biological hazards, chemical hazards and physical hazards. These microorganisms in foods that cause human to get sick are termed food borne pathogens, including bacterial, fungal, viral, and parasitic (protozoa and worms) organisms. These microorganisms and/or their toxins can cause biological hazards.

But the food borne illnesses caused by these microorganisms are sometimes incorrectly called food poisoning. There are two types of food poison caused by food borne pathogens: infections and intoxications. Infection food poison is caused by the ingestion of food containing a large amount of live pathogenic organisms. Intoxication food poison usually occurs by ingesting the food that is contaminated by the microorganisms or its toxins whether the microorganisms are alive or dead. The intoxications may be caused by improper storage conditions which allow the pathogens in food to grow and produce toxin. Subsequent processing of the food may destroy the microorganisms but not the toxin.

Mycotoxin is a metabolite produced by fungi growing in food or feed. At present, more than 200 varieties of mycotoxins have been identified. When humans or animals consumed food which was contaminated by mycotoxins, some multiple toxic symptoms will appear, such as hallucination, vomiting, haemorrhage, damage of central nervous system and even death. Fungal poisoning is common. The main characteristics of mycotoxin pathogenicity is the inability of human body to produce antibodies by the immune system. Several common fungal food poisonings are as follows: aflatoxin, sterigmatocystin, zearalenone, deoxynivalenol and so on.

2 COMMON MICROORGANISMS IN FOOD

Microorganisms in fresh produce include viruses (*Hepatitis* A virus and *Norovirus*); protozoa, such as *Cyclospora cayetanensis* and *Cryptosporidium parvum*; and bacteria such as *Aeromonas* (*A.*) *hydrophila*, *Bacillus* (*B.*) *cereus*, *Clostridium* (*Cl.*) spp., *Escherichia* (*E.*) *coli* O157 : H7, *Listeria* (*L.*) *monocytogenes*, *Salmonella* (*S.*) spp., *Shigella* (*Sh.*) spp., *Vibrio* (*V.*) *cholerae*, *Campylobacter* (*C.*) spp. and *Yersinia* (*Y.*) *enterocolitica*.

Species of *Pseudomonas*, *Micrococcus*, *Bacillus*, *Acinetobacter*, *Lactobacillus* and *Coryneform* are the prevalent bacteria in animal carcasses. Meanwhile, many samples also present *Coliforms*, *Staphylococcus* (*St.*) *aureus*, *Clostridium perfringens* and *Salmonellae*. Besides *Vibrio* species from seafood, other genera have been found in fishery products, including *Bacillus*, *Microbacterium*, *Micrococcus*, *Moraxella*, *Pseudomonas*, *Arthrobacter*, *Acinetobacter*, *Flavobacterium* and microorganisms designated as *Coryneforms*.

Parasites in foods usually cause illness when foods are consumed by human, such as *Cryptosporidium*, *Cyclospora*, *Giardia*, *Entamoeba*, *Toxoplasma*, *Sarcocystis*, *Isospora*, *Nematodes* and *Platyhelminthes*. In addition, food borne pathogens which can also spread via foods are *Hepatitis* A, *Norwalk* virus and *Norwalk-like* viruses, *Rotaviruses*, *Astroviruses*, *Enteroviruses*, *Parvoviruses*, *Adenoviruses* and *Coronaviruses*.

3 FOOD BORNE PATHOGENIC BACTERIA

Common food borne pathogenic bacteria include *Staphylococcus*, *Salmonella*, *Escherichia*, *Proteus*, *Listeria*, *Cl. botulinum*, *C. jejuni*, *V. cholerae*, *Vibrio parahaemolyticus*, *B. cereus*, and *Y. enterocolitica*. It is common to find that *Salmonella*, *E. coli* O157 : H7, *Staphylococcal*, and *Cl. botulinum* are the major pathogens contributing to outbreaks of food borne illness.

Salmonella is the most common and widely recognized food borne pathogen in the world, which widely occurs in families, schools, public catering units, etc. Usually food borne illness from *Salmonella* is linked to consumption of poultry; however, fresh produce has also been proven to be a frequent vehicle. In 2008 a large-scale outbreak of salmonellosis occurred in 43 states in the US and Canada, which involved 1442 illnesses that were linked to the consumption of hot peppers.

In 2006, a serious outbreak of severe illness was caused by *E. coli* O157 : H7-contaminated spinach in the US and Canada; 199 *E. coli* O157 : H7 cases with 3 deaths were reported in 26 states. 51% of these cases were hospitalized and 16% developed into acute renal failure. In May 2011, a food borne illness was caused by *E. coli* O104 : H4-contaminated fenugreek seed sprouts occurred in northern Germany which involved visitors from 14 European countries plus 7 persons (1 death) from North America. There were 3911 cases with 47 deaths and 777 patients developed hemolytic uremic syndrome (HUS).

4 MICROBIAL SOURCE

The microbiological safety of food remains a dynamic situation heavily influenced by multiple factors along the food chain from farm to fork. Microbial contamination can occur during any of the steps in the farm-to-consumer continuum (production, harvest, processing, wholesale storage, transportation or retailing and handling family) and this contamination can arise from environmental, animal or human sources.

Food is inevitably contaminated by different microorganisms in various links, such as production, processing, storage, transportation and sales. There are two main ways of microbial pollution: endogenous pollution and exogenous pollution. The food pollution phenomenon caused by microorganisms from animals and plants themselves is called endogenous pollution, also known as the first pollution. In the process of production, transportation, storage, marketing and consumption of food, microbial contamination of food is caused by water, air, human, animal, mechanical equipment and utensils, which is called exogenous pollution or second pollution. The microbes of contaminated food mainly come from soil, air, water, operators, animals and plants, processing equipment, packaging materials and so on. Environmental pollution, inadequate sterilization, improper storage and transportation methods, and inattention to sanitary operations are the main reasons for exceeding the standard of bacteria and pathogenic bacteria.

Soil is the natural living environment for a variety of human pathogens including *B. cereus*, *Cl. botulinum*, *Cl. perfringens*, *L. monocytogenes* and *Aeromonas*, but this profile of pathogens is broadened considerably by the addition of animal wastes into soil. The conditions at the growing location are major factors that affect the pathogen contamination of fresh produce. Fields that contain animal manure are more likely to be contaminated with enteric pathogens because of their ability to survive in soils for months or years. It has been reported that *E. coli* O157 : H7 and *Salmonella* may survive in soil for 7 to 25 weeks depending on the soil type, moisture level, temperature and source of contamination. Feces may naturally contain between 10^2 and 10^5 CFU/g *E. coli* and between 10^2 and 10^7 CFU/g *Salmonella* spp.; slurry between 10 and 10^4 CFU/g *E. coli* and *Yersinia* spp.; and manure between 10^2 and 10^7 CFU/g *Salmonella* spp. The manure of ruminants (cattle, sheep) and sewage are considered the main sources of *Salmonella* and *E. coli* O157 : H7. Also, *C. jejuni* is a normal member of the gastrointestinal micro flora of poultry, pigs and cattle. Since *L. monocytogenes* is widely distributed in nature (soil, decaying vegetation), the pathogen is a common contaminant of vegetables, especially root crops.

In addition, the pathogen populations relevant to food safety are not static. The previously unknown food borne pathogens, many of which are zoonotic, are constantly emerging.

5 FACTORS AFFECTING MICROBIAL GROWTH

The microenvironment in food products is changing constantly. These changes in the food affect the processes of microbial systems. The environmental conditions affect the survival of microorganisms. In turn, microorganisms also affect the environment.

A variety of intrinsic and extrinsic factors affect the metabolism and multiplication of microorganisms. The intrinsic or food related factors are pH, moisture content, water activity (Aw), oxidation-reduction potential, nutrients, and the possible presence of natural antimicrobial agents. The extrinsic or environment related factors include temperature, relative humidity, osmotic pressure, oxygen, inhibitors, light, and the type and quantity of microorganisms present in food.

Most of the bacteria, especially the pathogenic bacteria, grow best and produce most toxins when pH is 7.0 (6.6 ~ 7.5), and a few bacteria can also grow when pH is below 4.0. Bacteria require stricter pH than yeast and mold, and pathogenic bacteria are more severe. Neutral food (pH 6.6~7.5) is one of the most common pathogenic bacteria survival routes.

Microbial requirements for water are generally expressed in terms of water activity. Most fresh foods have a water activity of 0.99, and each microbial growth and toxin production requires a minimum water activity. For pathogenic bacteria, *St. aureus* can grow when A_w is below 0.86, but *Cl. botulinum* cannot grow when A_w is lower than 0.94, and mold can also grow when A_w is below 0.8. The less moisture in food, the lower the ability of microorganisms to produce toxins.

Nutrients in food can provide water, carbon sources, nitrogen sources, energy, vitamins, minerals and related growth factors for microorganisms. The more nutritious ingredients in foods, the faster the pathogenic bacteria grow and reproduce, and the more dangerous they are.

Some foods have natural antimicrobial substances, such as antibacterial oils (eugenol, allicin, cinnamaldehyde, mustard oil). There are also many antimicrobial components, such as lactoferrin, lectin, the peroxidase system in milk, lysozyme in eggs, thioglycoside in cruciferous plants and so on.

Most pathogenic bacteria belong to mesophilic microorganisms. To some extent, cold storage can inhibit the growth of pathogenic bacteria. When food is frozen, some microbial groups can be effectively killed. When food is heated, the key protein, nucleic acid and enzyme system of the bacteria can be destroyed directly, causing the death of the bacteria.

Adjusting the gas composition and proportion of food environment, on the one hand, can prevent the growth of spoilage microorganisms and inhibit food spoilage. On the other hand, it can control the growth and reproduction of pathogenic bacteria by inhibiting the growth of microorganisms.

6 MICROBIAL CONTROL

Microbial control is the use of physical, chemical and biological methods to prevent microbial contamination, and kill or inhibit the growth and reproduction of microorganisms. Physical methods can affect the chemical composition and metabolism of microbial growth, so this kind of methods can be used to suppress or kill microorganisms. The physical methods for controlling microorganisms are mainly heating, radiation, drying, ultrasound, filtration, cleaning, microwave, high pressure, ohmic heating, far infrared, low temperature, and so on. The commonly used high-temperature sterilization methods are dry heat sterilization and damp heat sterilization. The dry heat sterilization includes burning and hot-air sterilization. Damp heat sterilization includes pasteurization, boiling, circulating steam disinfection, intermittent sterilization, normal autoclaving, continuous autoclaving sterilization and so on. Bacterial species, number of bacteria, composition of food, heating temperature and time, all affect the efficacy of thermal sterilization.

Chemical methods to control microorganisms include salt, sugar, wine, vinegar and preservatives. Preservatives include benzoic acid and its salts, sorbic acid and its salts, sodium diacetate, o-phenylphenol, hydrogen peroxide, sulfites, nitrate and nitrite, etc. Action mechanism of preservatives: (1) To destroy the cell wall and cell membrane of a microorganism. For example, surfactant can depolymerize the cell wall of Gram negative bacteria. Phenols and alcohols can lead to cell membrane structure disorder and interfere with its normal function, and make the small molecular metabolites overflow the cells. (2) To cause microbial protein denaturation or coagulation. Acids and alkalis, alcohols, aldehydes, dyes, heavy metal salts and oxidants can cause microbial protein denaturation or coagulation. For example, ethanol can change the configuration of bacterial protein and disrupt the folding mode of polypeptide chain, resulting in protein denaturation. (3) Change the structure of nucleic acids and inhibit the synthesis of nucleic acids. Some aldehydes, dyes and alkylating agents play bactericidal and bacteriostatic roles by affecting the biosynthesis and function of nucleic acids. For example, formaldehyde can bind to the amino group on the base ring of microbial nucleic acids.

Biological methods for microbial control currently include biological preservatives (Nisin, Natamycin), application of bacteriophages, antagonistic bacteria, and a combination of antagonistic bacteria with bacteriophages.

NOTES

1. Some microorganisms that do not directly cause changes in food may alter the flora so that spoilage organisms can grow.

2. The conditions at the growing location are major factors that affect the pathogen contamination of fresh produce.

3. Only when food is frozen to form ice crystals can some microbial groups be effectively killed.

4. When food is heated, the key protein, nucleic acid and enzyme system of the bacteria can be destroyed directly, causing the death of the bacteria.

🔊 Lesson 7

Vocabulary

/ Words /

parasitic [ˌpærəˈsɪtɪk] adj. 寄生的, 由寄生虫引起的

protozoa [ˌproʊtəˈzoʊə] n. 原生动物

contaminate [kənˈtæmɪneɪt] vt. 污染

feed [fi:d] n. 饲料

consume [kən'su:m] vt. 消耗, 消费, 吃光

initiate [ɪ'nɪʃieɪt] vt. 开始, 发起

symptom ['sɪmptəm] n. 症状

hallucination [həˌlu:sɪ'neɪʃn] n. 幻想, 错觉

vomiting ['vɑ:mɪtɪŋ] n. 呕吐

hacmorrhagc ['hcmərɪdʒ] n. 出血

pathogenicity [pæθədʒɪ'nɪsɪtɪ] n. 病原性, 致病性

immune [ɪ'mju:n] adj. 免疫的, 有免疫力的

aflatoxin [ˌæfləˈtɑksɪn] n. 黄曲霉毒素

sterigmatocystin [stɪrɪgmətoʊ'sɪstɪn] n. 杂色曲霉毒素

zearalenone [ziːəˈrælənəʊn] n. 玉米赤霉烯酮

deoxynivalenol [dɪɒksɪnɪvəˈlinɒl] n. 脱氧雪腐镰刀菌烯醇

hepatitis [ˌhepəˈtaɪtɪs] n. 肝炎

Norovirus [nɔːroʊvaɪrəs] 诺如病毒

Salmonella [ˌsælməˈnelə] n. 沙门氏菌

Clostridium [klɒsˈtrɪdɪəm] n. 梭菌, 梭菌属

Shigella [ʃɪˈgelə] n. 志贺氏（杆）菌

Campylobacter [ˈkæmpɪloˌbæktər] n. 弯曲杆菌

endogenous [enˈdɑ:dʒənəs] adj. 内长的, 内生的

exogenous [ekˈsɑ:dʒənəs] adj. 外成的, 外生的

utensil [ju:ˈtensl] n. 器具

sanitary [ˈsænəteri] adj. 清洁的, 卫生的

enteric [enˈterɪk] adj. 肠的

manure [məˈnʊr] n. 肥料, 粪便

ruminant [ˈru:mɪnənt] n. 反刍动物

decay [dɪˈkeɪ] vt. & vi. （使）腐烂, 腐朽

vegetation [ˌvedʒəˈteɪʃn] n. 植物

zoonotic [ˌzəʊəˈnɒtɪk] 动物传染病, 人畜共患病

intrinsic [ɪnˈtrɪnzɪk] adj. 固有的, 内在的

extrinsic [eksˈtrɪnzɪk] adj. 非本质的, 外在的

metabolism [məˈtæbəlɪzəm] n. 新陈代谢

multiplication [ˌmʌltɪplɪˈkeɪʃn] n. 增加, 增殖

antimicrobial [ˌæntɪmaɪˈkrəʊbɪəl] adj. 抗菌的

reproduction [ˌri:prəˈdʌkʃn] n. 繁殖, 生殖

suppress [səˈpres] vt. 压制, 阻止…的生长

mesophile [ˈmezəˌfaɪl] adj. （细菌）嗜温的, 适温生物

radiation [ˌreɪdiˈeɪʃn] n. 辐射

sulfite [ˈsʌlˌfaɪt] n. 亚硫酸盐

Pseudomonas [sudəˈmoʊnəz] n. 假单胞菌

Micrococcus [ˌmaɪkrəˈkɒkəs] n. 微球菌, 球状细菌

Lactobacillus [ˌlæktoʊbəˈsɪləs] n. 乳酸菌

Coryneform [kəˈrɪnəˌfɔrm] adj. 棒状杆菌的, 似棒状杆菌的

Coliforms [kəlɪˈfɔ:mz] 大肠杆菌类

Vibrio [ˈvɪbrɪˌoʊ] n. 弧菌

Microbacterium [ˌmaɪkrɒˈbæktɪərɪm] 细杆菌属, 小细菌属

parasite [ˈperəˌsaɪt] n. 寄生虫

Moraxella [ˌmɔːrækˈselə] 莫氏杆菌属

Flavobacterium [fleɪvɒˈbæktɪərɪm] n. 黄质菌属, 产黄菌属

Giardia [dʒiːˈɑːdiːə] n. 贾第鞭毛虫属

Entamoeba [ˌentəˈmiːbə] n. 内阿米巴属

Toxoplasma [tɒksə'plæzmə] n. 弓形体，弓浆虫

Sarcocystis [sɑːkəsɪs'tɪs] n. 肉孢子虫

Isospora [aisəu'spɔːrə] n. 等孢子球虫属

Nematode ['nemətəʊd] 线虫类

Platyhelminthes [plæti:'helmɪnθs] 扁形动物

Proteus ['prəʊtiːəs] 变形杆菌

poultry ['pəʊltri] n. 家禽

spinach ['spɪnɪtʃ] n. 菠菜

fenugreek ['fenjugriːk] n. 葫芦巴（一年生豆科植物）

sprout [spraʊt] n. 新芽，嫩芽

continuum [kən'tɪnjuəm] n. 连续体，连续统一体

wholesale ['həʊlseɪl] n. 批发，大规模买卖

inevitably [ɪn'evɪtəbli] adv. 不可避免地

nitrate ['naɪtreɪt] n. 硝酸盐

nitrite ['naɪtraɪt] n. 亚硝酸盐

depolymerize [dɪ'pɒlɪməraɪz] v. （使）解聚

phenol ['fiːnɔːl] n. 酚类

alcohol ['ælkəhɔːl] n. 醇类，含酒精的饮料

overflow [oʊvər'floʊ] vt. & vi. 溢出，淹没

denaturation [diːˌneɪtʃə'reɪʃən] n. 改变本性，变性

coagulation [koʊˌægju'leɪʃn] n. 凝结，凝结物

alkali ['ælkəlaɪ] n. 碱

aldehyde ['ældəˌhaɪd] n. 醛

oxidant ['ɑːksɪdənt] n. 氧化剂

ethanol ['eθənɔːl] n. 乙醇

configuration [kənˌfɪgjə'reɪʃn] n. 布局，构造；[化]（分子中原子的）组态，排列

disrupt [dɪs'rʌpt] vt. 使混乱，使分裂，破坏

polypeptide [pɒlɪ'peptaɪd] n. 多肽

bactericidal [bækˌtɪərɪ'saɪdl] adj. 杀菌的

bacteriostatic [bækˌtɪrɪəs'tætɪk] adj. 细菌抑制的，阻止细菌繁殖法的

formaldehyde [fɔːr'mældɪhaɪd] n. 甲醛，福尔马林

natamycin [neɪtə'maɪsɪn] n. 纳他霉素

bacteriophage [bæk'tɪərɪəfeɪdʒ] n. 噬菌体

antagonistic [ænˌtægə'nɪstɪk] adj. 敌对的，对抗性的

/ Phrases /

food spoilage 食品腐败
food borne illness 食源性疾病
food borne pathogens 食源性致病菌
food poisonings 食物中毒
infection and intoxication 感染型和毒素型
fresh produce 新鲜农产品，生鲜农产品
Cyclospora cayetanensis 环孢子虫
Cryptosporidium parvum 小球隐孢子虫
Aeromonas hydrophila 嗜水气单胞菌
Bacillus cereus 蜡样芽孢杆菌
Listeria monocytogenes 单核细胞增生李斯特菌
Vibrio cholerae 霍乱弧菌
Yersinia enterocolitica 结肠炎耶尔森杆菌
Clostridium botulinum 肉毒梭状芽孢杆菌
Campylobacter jejuni 空肠弯曲杆菌
Vibrio parahaemolyticus 副溶血性弧菌
hot peppers 辣椒
acute renal failure 急性肾功能衰竭
hemolytic uremic syndrome（HUS）溶血性尿毒症综合征

  Clostridium perfringens 产气荚膜梭状芽孢杆菌
  Acinetobacter 不动细菌属
  Staphylococcus aureus 金黄色葡萄球菌
  Arthrobacter 节细菌属
  Norwalk virus 诺瓦克病毒
  Norwalk-like virus 类诺瓦克病毒
  Rotaviruses 轮状病毒
  Astroviruses 星状病毒
  Enteroviruses 肠病毒
  Parvoviruses 细小病毒科
  Adenoviruses 腺病毒
  Coronaviruses 冠状病毒
  animal wastes 动物粪便
  root crops 块根植物
  relative humidity 相对湿度
  osmotic pressure 渗透压
  ohmic heating 欧姆加热
  far infrared 远红外
  antagonistic bacteria 拮抗菌
  hot-air sterilization 热空气消毒法
  circulating steam disinfection 循环蒸汽消毒
  intermittent sterilization 间歇灭菌
  normal autoclaving 高压蒸汽灭菌法
  continuous autoclaving sterilization 连续高压灭菌
  gram negative bacteria 革兰氏阴性菌
  alkylating agent 烷化剂

Exercises

I. Write true or false for the following statements according to the passage.

1. (　) Microorganisms may cause food spoilage or food borne illness.

2. (　) Food borne illnesses are only caused by microorganisms.

3. (　) These microorganisms in foods that cause human to get sick are termed food borne pathogens, including bacterial, fungal, viral, and parasitic (protozoa and worms) organisms.

4. (　) There is one type of disease caused by food borne pathogens: infection which results from ingestion of live pathogenic organisms.

5. (　) Subsequent processing of the food may destroy the microorganisms and the toxin.

6. (　) Main characteristics of mycotoxin pathogenicity are that the body produces antibodies to mycotoxins and immune to them.

7. (　) There are two main ways of microbial pollution: endogenous pollution and exogenous pollution.

8. (　) The environmental conditions determine the species of existed microorganisms, in turn, the microorganisms also affect the environment.

9. (　) The intrinsic or food related factors which affect the metabolism and multiplication of microorganisms are pH, moisture content, water activity, oxidation-reduction potential, nutrients, and the possible presence of natural antimicrobial agents.

10. (　) Microbial control is the use of physical, chemical and biological methods to kill the microorganisms in foods.

II. Answer the following questions according to the passage.

1. What is the definition of food borne pathogens?
2. What is infectious food poisoning?
3. Where do the microorganisms that contaminate foods come from?
4. What are the extrinsic factors affecting the growth of microorganisms?
5. How to inhibit/control the growth of microorganisms?

III. Fill in the blanks according to the passage.

1. Food borne pathogens include＿＿＿＿，＿＿＿＿，＿＿＿＿, and＿＿＿＿ organisms.
2. ＿＿＿＿ usually occur when the microorganisms and their toxins contaminated food and they are consumed, no matter the microorganisms are dead or alive.
3. There are two main ways of microbial pollution: ＿＿＿＿ pollution and＿＿＿＿ pollution.
4. A variety of＿＿＿＿ and＿＿＿＿ factors affect the metabolism and multiplication of microorganisms.
5. Methods of chemical to control of microorganisms include＿＿＿＿，＿＿＿＿，＿＿＿＿, ＿＿＿＿ and＿＿＿＿.

IV. Translate the following words and expressions into Chinese.

food borne pathogens   *Listeria monocytogenes*
Bacillus cereus   *Staphylococcus aureus*
antagonistic bacteria   *Escherichia coli*
Yersinia enterocolitica   relative humidity
Clostridium botulinum   osmotic pressure

V. Translate the following expressions into English.

1. 多种内在和外在因素影响微生物的代谢和增殖，其中内在或与食物有关的因素有pH、水分含量、水分活度、氧化还原电位、营养物质和可能存在的天然抗菌物质。外部或环境相关因素包括温度、相对湿度、渗透压、氧气、抑制剂、光以及食物中存在的微生物的类型和数量。

2. 物理因素能影响微生物生长的化学组成和新陈代谢，因此可以用物理方法抑制或杀死微生物，控制微生物的物理方法主要有加热、辐射、干燥、超声波、过滤、清洗、微波、高压、欧姆加热、远红外、低温等。

# 参考译文

第 7 课　食品安全与微生物影响因素

1　微生物与食品安全

微生物可引起食品腐败或食源性疾病。食品腐败是指食品在气味、颜色、口感、质地和外观等方面产生不良变化,甚至失去其食用价值。一些微生物不直接引起食品的变化,可能会改变微生物区系,使腐败微生物得以生长。

食源性疾病通常由微生物或食物中的有毒物质引起,包括生物危害、化学危害和物理危害。这些导致人类生病的食物中的微生物被称为食源性致病菌,包括细菌、真菌、病毒和寄生虫(原生动物和蠕虫)。这些食源性致病菌和/或它们的毒素会造成生物危害。

由这些微生物引起的食源性疾病有时被不恰当地称为食物中毒。由食源性致病菌引起的食物中毒有两种:感染型和毒素型。感染型食物中毒是由于摄入的食物中含有大量活的致病菌而引起的。毒素型食物中毒通常是由于微生物及其毒素污染食物并被食用而引起,无论微生物是死的还是活的都可能引起毒素型食物中毒。由于储存不当,食源性致病菌在食品中生长并产生毒素从而导致毒素型食物中毒。食物的后续加工可能会破坏微生物,但不会破坏毒素。

真菌毒素是真菌在食品或饲料里生长所产生的代谢产物,目前已知的真菌毒素有 200 多种。人或动物摄入被真菌毒素污染的农、畜产品可引发多种中毒症状,如致幻、催吐、出血症、中枢神经受损,甚至死亡。真菌中毒现象普遍存在。真菌毒素致病的主要特点:机体对真菌毒素不能产生抗体,也不能免疫。几种常见的与食物中毒有关的真菌如下:黄曲霉毒素、杂色曲霉毒素、玉米赤霉烯酮以及脱氧雪腐镰刀菌烯醇等。

2　食品中常见的微生物

生鲜农产品中常见的微生物包括:病毒(甲型肝炎病毒和诺如病毒);原生动物,如环孢子虫、小球隐孢子虫等;以及细菌,如嗜水气单胞菌、蜡状芽孢杆菌、某些梭状芽孢杆菌、大肠杆菌 O157︰H7、单核细胞增生李斯特菌、某些沙门氏菌、某些志贺氏菌、霍乱弧菌、某些弯曲杆菌和小肠结肠炎耶尔森杆菌。

动物尸体上常见的细菌有假单胞菌、微球菌、芽孢杆菌、不动杆菌、乳酸杆菌和棒状菌等,同时许多样品中也含有大肠杆菌类、金黄色葡萄球菌、产气荚膜梭菌和沙门氏菌。除来自海产品的弧菌外,渔业产品中还发现其他几属的微生物,如芽孢杆菌属、微杆菌属、微球菌属、莫拉氏菌属、假单胞菌属、节杆菌属、不动杆菌属、黄杆菌属和类棒状杆菌属的生物。

食物中的寄生虫,如隐孢子虫、环孢子虫、贾第鞭毛虫、内阿米巴虫、弓形虫、肉孢子虫、等孢子球虫、线虫、扁形动物等，通常在人类食用时引起疾病。此外，甲型肝炎病毒、诺瓦克病毒和类诺瓦克病毒、轮状病毒、星形病毒、肠道病毒、细小病毒、腺病毒和冠状病毒等食源性致病菌也可以通过食物传播。

3　食源性致病菌

常见的食源性致病菌包括：葡萄球菌、沙门氏菌、大肠杆菌、变形杆菌、单核细胞增生李斯特菌、肉毒梭状芽孢杆菌、空肠弯曲菌、霍乱弧菌、副溶血性弧菌、蜡样芽孢杆菌、小肠结肠炎耶尔森杆菌。通常沙门氏菌、大肠杆菌O157：H7、葡萄球菌、肉毒梭状芽孢杆菌是导致食源性疾病爆发的主要致病菌。

沙门氏菌是世界上最常见、最普遍的食源性致病菌，广泛存在于家庭、学校、公共餐饮单位等。通常由沙门氏菌引起的食物传播疾病与食用禽类有关，但是生鲜农产品也是其主要来源。2008年，美国和加拿大的43个州爆发了大规模的沙门氏菌病，其中1442例疾病与食用辣椒有关。

2006年，在美国和加拿大的26个州出现199例由大肠杆菌O157：H7污染的菠菜引起的严重疾病，其中3例死亡。199例中51%的病例住院，16%的病例发生急性肾功能衰竭。2011年5月，在德国北部发生了由大肠杆菌O104：H4污染的葫芦巴种子芽引起的大规模疾病，致使来自14个欧洲国家的游客和来自北美的7人（1人死亡）全部染病。本次疾病共导致3911起病例，其中47人死亡，777名患者出现溶血性尿毒症综合征（HUS）。

4　微生物来源

食品的生物安全是一个动态状态，主要受从农田到餐桌食物链上的多重因素影响。微生物污染可能发生在农场到消费者的任何环节（生产、收获、加工、批发储存、运输、零售和家庭处理），这种污染可能来自环境、动物或人类。

食品在生产、加工、贮藏、运输和销售等各个环节，不可避免地会遭受不同微生物的污染。微生物污染途径主要分为内源性污染和外源性污染两种。凡是动植物体在生活过程中，由于自身带有的微生物而对食品造成的污染现象，称为内源性污染，也称为第一次污染。食品在生产加工、运输、贮藏、销售及食用过程中，通过水、空气、人、动物、机械设备及其用具等而造成微生物对食品的污染现象，称为外源性污染，也称为第二次污染。污染食品的微生物主要来源于土壤、空气、水、操作人员、动植物、加工设备、包装材料等方面。环境污染、灭菌不当、贮存运输方法不当、卫生操作不规范是造成细菌和致病菌超标的主要原因。

土壤是人类致病菌的天然生存环境，如蜡状芽孢杆菌、肉毒梭状芽孢杆菌、产气荚膜梭菌、单核细胞增生李斯特菌和气单胞菌属，但随着动物粪便进入土壤，病原菌的分布范围也会扩大。生鲜农产品的生长条件是影响微生物污染的主要因素。含有动物粪便的农田更容易受到肠道病原菌的污染，因为它们能够在土壤中存活数月或数年。据报道，由于土壤类型、湿度、温度和污染源的不同，大肠杆菌O157：H7和沙门氏菌可在土壤中存活7~25周。粪便中含有

$10^2 \sim 10^5$ CFU/g 大肠杆菌，$10^2 \sim 10^7$ CFU/g 沙门氏菌；泥浆中含有 $10 \sim 10^4$ CFU/g 大肠杆菌和耶尔森菌，肥料中含有 $10^2 \sim 10^7$ CFU/g 沙门氏菌。反刍动物（牛、羊）的粪便和排泄物被认为是沙门氏菌和大肠杆菌 O157：H7 的主要来源。此外，空肠弯曲菌也是家禽、猪和牛胃肠道微生物区系的正常成员。由于单核细胞增生李斯特菌广泛分布于自然界（土壤、腐烂植被），该病原体是蔬菜，尤其是根作物的常见污染物。

此外，与食品安全有关的致病菌种群也不是一成不变的，以往未知的食源性致病菌正在不断出现，其中许多会导致人畜共患病。

5 影响微生物生长的因素

食品中的微环境是不断变化的，这些变化影响着微生物区系的变化。环境条件决定了微生物的种类，反过来，微生物又影响环境。

多种内在和外在因素影响微生物的代谢和增殖，其中内在或与食物有关的因素有 pH、水分含量、水分活度、氧化还原电位、营养物质和可能存在的天然抗菌物质。外部或环境相关因素包括温度、相对湿度、渗透压、氧气、抑制剂、光以及食物中存在的微生物的类型和数量。

大多数细菌，尤其是致病菌性细菌，在 pH 7.0（6.6~7.5）时生长和产毒最好，个别细菌 pH 4.0 以下也能生长。细菌对 pH 要求比酵母和霉菌苛刻，致病性细菌更加苛刻，近中性食品（pH 6.6~7.5）是致病菌最常见的传播途径之一。

微生物对水分的要求一般用水分活度表示。大多数新鲜食品的水分活度都在 0.99，每种微生物生长和产毒都需要一个最低的水分活度。对于致病菌来说，金黄色葡萄球菌水分活度低于 0.86 还能够生长；而肉毒梭状芽孢杆菌在水分活度低于 0.94 时就不能生长；霉菌在水分活度低于 0.8 时也能够生长。食品中水分越少，微生物的产毒能力也就越低。

食品中的营养成分可为微生物提供水分、碳源、氮源、能量、维生素、矿物质和相关的生长因子等。食品中营养成分越丰富，致病性细菌的生长繁殖就越快，危险性也增加。

有些食品具有天然的抑菌物质，如具有抗菌功能的油（丁子香酚、大蒜素、肉桂醛、芥子油）。牛乳中也有多种抗菌成分，如乳铁蛋白、凝集素、过氧化酶系统。鸡蛋中的溶菌酶，十字花科植物的硫代葡萄糖苷等。

大多数致病性细菌属于嗜温性微生物，冷藏一定程度上可抑制致病性细菌的生长。只有食品在冷冻形成冰晶时，才能有效杀灭一些微生物类群。食品在热加工时，可使菌体关键蛋白质、核酸、酶系统都直接受到破坏作用，致使菌体死亡。

调整食品环境的气体组成及比例，一方面可以防止腐败微生物的生长，抑制食品腐败变质；另一方面通过抑制微生物的生长也控制了致病性细菌的生长和繁殖。

6 微生物控制

微生物控制就是采用物理、化学和生物学的方法防止微生物污染、杀灭或抑制微生物的生长繁殖。物理因素能影响微生物生长的化学组成和新陈代谢，因此可以用物理方法抑制或杀死微生物，控制微生物的物理方法主要有加热、辐射、干燥、超声波、过滤、清洗、微波、高压、欧姆加热、远红外、低温等。常用的高温灭菌方法有干热灭菌法和湿热灭菌法。干热灭菌法包括灼烧法和热空气消毒法。湿热灭菌法包括巴氏消毒法、煮沸法、流通蒸汽消毒法、间歇灭菌法、常规高压蒸汽灭菌法、连续加压灭菌法等。菌种、菌体数量、食品的成分、加热的温度和时间都影响热杀菌的效果。

化学控制微生物的方法主要有盐、糖、酒、醋和防腐剂等。防腐剂主要包括苯甲酸及其盐类、山梨酸及其盐类、双乙酸钠、邻苯基苯酚、过氧化氢、亚硫酸盐、硝酸盐和亚硝酸盐等。防腐剂的作用机制：（1）破坏微生物的细胞壁、细胞膜。如表面活性剂可使革兰氏阴性菌的细胞壁解聚。酚类和醇类可导致微生物细胞膜结构紊乱并干扰其正常功能，使细胞的小分子代谢物质溢出胞外。（2）引起微生物菌体蛋白质变性或凝固。酸碱、醇类、醛类、染料、重金属盐和氧化剂等消毒防腐剂有此作用。如乙醇可引起菌体蛋白质构型改变而扰乱多肽链的折叠方式，造成蛋白质变性。（3）改变核酸结构、抑制核酸合成。部分醛类、染料和烷化剂通过影响核酸的生物合成和功能发挥杀菌、抑菌作用。如甲醛可与微生物核酸碱基环上的氨基结合。

目前用于微生物控制的生物方法包括：生物防腐剂（乳酸链球菌素，纳他霉素），噬菌体、拮抗细菌和噬菌体与拮抗菌的结合方法。

备注：
1. 一些不会直接导致食物变化的微生物可能会改变菌群，从而使腐败微生物生长。
2. 生长地点的条件是影响新鲜农产品病原体污染的主要因素。
3. 只有当食物冷冻形成冰晶时，才能有效杀死某些微生物群。
4. 当食物被加热时，细菌的关键蛋白质、核酸和酶系统会被直接破坏，导致细菌死亡。

LESSON 8

FOOD QUALITY MANAGEMENT

Food quality management (FQM) focuses on consumer-driven quality management in food production systems using a product-based approach. FQM integrates organizational and technological aspects of food product quality into one techno-managerial concept and presents an integrated view of how quality management is to be situated in a chain-oriented approach. Quality management is concerned with controlling activities with the aim of ensuring that products and services are fit for their purpose and meet specifications.

"Food quality management" is defined here as all of the activities of the overall management function that determine the quality policy, objectives, and responsibilities and that implement them by means of quality planning (QP), quality control (QC), quality assurance (QA), and quality improvement (QI) within the quality system.

The implementation of quality management depends on organizational factors such as the size of the organization, the type of suppliers and customers, the degree of automation, the type of products, QA requirements, and most importantly, top management's commitment.

1 PRINCIPLES OF QUALITY MANAGEMENT

Organizations that focus on quality rely on the same basic management principles for success. Table 3 lists seven principles of quality management.

Table 3   Principles of Quality Management

	Quality principles	Description
1	Customer first	Sustained success is achieved when an organization attracts and retains the confidence of customers and other interested parties. Understanding current and future needs of customers and other interested parties contributes to sustained success of the organization.
2	Leadership	Organizations succeed when leaders establish and maintain the internal environment in which employees can become fully involved in achieving the organization's unified objectives.

Continued table

	Quality principles	Description
3	Employee engagement	Organizations succeed by retaining competent employees, encouraging continuous enhancement of their knowledge and skills, and empowering them, encouraging engagement, and recognizing achievements.
4	Process approach	The quality management system consists of interrelated processes. Understanding how results are produced by which to enable an organization to optimize the system and its performance.
5	Improvement	Improvement is essential for an organization to maintain current levels of performance, to react to changes in its internal and external conditions, and to create new opportunities.
6	Evidence-based decision-making	Decision-making can be a complex process, and it always involves some uncertainty. Facts, evidence, and data analysis lead to greater objectivity and confidence in decision-making.
7	Relationship management	Interested parties influence the performance of an organization. Sustained success is more likely to be achieved when the organization manages relationships with all of its interested parties to optimize their impact on its performance.

2 FUNCTIONS OF QUALITY MANAGEMENT

Management is creative problem-solving. This creative problem-solving is accomplished through five functions of management: quality policy and strategy (QP&S), quality design (QD), quality control (QC), quality assurance (QA), and quality improvement (QI) (Figure 3). QI helps companies to improve current business processes, whereas QA satisfies regulatory requirements and improves food quality safety. Each FQM function consists of diverse methods, tools, and techniques.

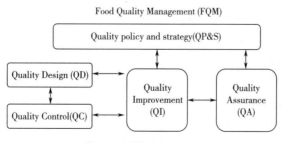

Figure 3 FQM function

2.1 Quality Policy and Strategy (QP&S)

QP&S ensures that a quality management system (QMS) is included in a company's long-term business strategy and helps a company to take appropriate actions and allocate resources to achieve those goals.

Quality policy refers to the overall quality purpose and direction of the organization as officially issued by the top management of the organization. Quality management principles can be used as the basis for formulating quality policy. Quality policy is the guiding principle of the enterprise's quality behavior, which reflects the quality consciousness of the top managers of the enterprise. It also reflects the aim and culture of quality management. In a sense, the quality policy is the concept of quality management.

The increasingly competitive market and mounting demand for high-quality products from consumers compel food companies to assimilate quality values in their strategic document. Total quality management (TQM), quality cost analysis (QCA), and strategy analysis (SA) are commonly used methods. TQM is a management system that includes the core values of the organization, tools, and techniques. The objective of QCA is to measure the effectiveness of the firm's operation at all levels and minimize the production costs. Similarly, companies use SA as a process to establish the most appropriate management system with the least resistance to achieve business results.

2.2 Quality Design (QD)

QD is a method to translate the voice of the customer into the technical requirements of the products and processes with the help of specific techniques so that the final quality meets or exceeds customers' expectations. High quality can be achieved only through a robust QD by targeting the source of the production process.

Managers and planners can help ensure good care and prevent problems from arising by designing quality into every aspect of a program, including its mission and objectives, allocation of resources, and development of standards and guidelines.

QD begins with defining the organization's mission, including its purpose, value, objectives, and clients, with an eye to quality. This is the first step both for creating a new product and for redesigning an existing product. To develop realistic objectives concerning quality, managers must clearly assess the level of quality that can be achieved with available resources; the institution's strengths and weaknesses, including current program performance and quality; the client population, including how clients themselves perceive quality; and the political and social climate. Objectives foster quality best when they focus on meeting the client's needs.

In QD, the participation of front-line service providers, producers, field supervisors, and clients is crucial. Top decision-makers rarely have direct experience with day-to-day produce-and-service delivery. Without advice from providers and clients, intended improvements may not be meaningful to the staff who must implement them and may not attract clients or meet people's needs.

2.3 Quality Control (QC)

QC plays a significant role in the food sector because there is a huge difference between food products and biological raw materials. The basic objective of QC is to control variation at a tolerable level by taking corrective actions. Statistical and non-statistical tools and techniques have been developed and implemented to measure, analyze, and control variation in food products. Statistical process control, acceptance sampling, and visual inspection are extensively used in both food and nonfood sectors.

QC ensures that a program's activities be conducted as designed. QC activities also may uncover flaws in design and thus point out changes that could improve quality. For foods, the main objective of QC is to ensure that all of the processes (from farm to table) are controlled. QC includes day-to-day supervision and monitoring to confirm that activities are proceeding as planned and staff members are following guidelines. It also includes periodic evaluations that measure progress toward program objectives. Good QC requires that programs develop and maintain measurable indicators of quality, timely data collection and analysis, and effective supervision.

2.4 Quality Assurance (QA)

QA plays a significant role in food sector. It provides a guarantee that all of the quality obligations such as food safety and reliability are met through the establishment of a standard organizational structure, responsibilities, processes, and procedures.

QA is a method of preventing mistakes or defects in manufactured products and avoiding problems when delivering solutions or services to customers, which the International Organization for Standardization (ISO) 9000 defines as "part of quality management focused on providing confidence that quality requirements will be fulfilled." This defect prevention in QA differs subtly from defect detection and rejection in QC and has been referred to as a "shift left," as it focuses on quality earlier in the process.

QA includes two principles: "fit for purpose" (the product should be suitable for the intended purpose) and "right first time" (mistakes should be eliminated). QA includes management of the quality of raw materials, assemblies, products and components, services related to production and management, production, and inspection processes. The two principles also manifest prior to the background stage of developing (engineering) a novel technical product: the task of engineering is to make it work once, while the task of QA is to make it work all of the time.

Several QA systems have been developed to fit the needs of the food sector, such as hazard analysis critical control points (HACCP), the ISO, the International Food System, the British Retail Consortium, and Quality System (QS).

2.4.1 Good Agricultural Practice

Good agricultural practice (GAP) is a specific method and system which, when applied to agriculture, creates food for consumers or further processing what is safe and wholesome. While there are numerous competing definitions of what methods constitute good agricultural practice, there are several broadly accepted schemes that producers can adhere to. GAP points out farming methods that reduce the likelihood of contaminating primary produce. Implementation of practices address water quality, manure and compost use, worker health and hygiene and contamination from wild life, domestic animals and livestock.

2.4.2 Good Manufacturing Practice

Good manufacturing practice (GMP) is a set of operational requirements to ensure that processors produce safe and quality food. The FDA and other regulatory bodies worldwide highly recommend a GMP to ensure that products are consistently produced and controlled to quality standards. GMP specification covers the minimum sanitary and manufacturing practices that are a prerequisite foundation to further implement other food safety management programs such as HACCP, ISO 22000, IFS and SQF.

2.4.2.1 Sanitary Operations

Factory buildings, fixed equipment, and other physical facilities shall be maintained in a sanitary condition and shall be kept in repair sufficiently to prevent food from becoming adulterated meant by the act. Cleaning and sanitizing of utensils and equipment shall be conducted in a manner that protects against contamination of food, food-contact surfaces, or food-packaging materials. And no pests shall be allowed in any area of a food factory.

Cleaning compounds and sanitizing agents used in cleaning and sanitizing procedures shall be free from undesirable microorganisms and shall be safe and adequate under the conditions of use. Toxic cleaning compounds, sanitizing agents, and pesticide chemicals shall be identified, held, and stored in a manner that protects against contamination of food, food-contact surfaces, or food-packaging materials.

2.4.2.2 Processes and Controls

All operations in the receiving, inspecting, transporting, segregating, preparing, manufacturing, packaging, and storing of food shall be conducted in accordance with adequate principles of sanitation. Appropriate quality control operations shall be employed to ensure that food is suitable for human consumption and that food-packaging materials are safe and suitable. Overall sanitation of the plant shall be under the supervision of one or more competent individuals assigned responsible for this function. All reasonable precautions shall be taken to ensure that production procedures do not lead to contamination from any source. Chemical, microbial, or extraneous-material testing procedures shall be used when necessary to identify sanitation failures or possible food contamination. All food that has become contaminated to the extent that it is adulterated within the provisions of the act shall be recalled, or if permissible, treated or processed to eliminate the contamination.

2.4.3 Sanitation Standard Operating Procedures

Sanitation standard operating procedures (SSOP) is the common name given to the procedures of sanitation in food production plants. It is considered as one of the prerequisite programs of HACCP.

SSOP can be very simple to extremely intricate depending on its focus. Food industry equipment should be designed in a way that is convenient for sanitation and cleaning; however, some automated processing equipment necessary for cleaning is difficult to clean. An individual SSOP should include: specify the name of the equipment to be cleaned or an area of the affected equipment; prepare necessary tools to clean the equipment to be cleaned; how to disassemble the whole equipment or part of it; and the method of cleaning and sanitizing.

2.4.3.1 Sanitation Controls

The SSOP shall address: safety of the water that comes into contact with food or food contact surfaces or that is used in the manufacture of ice; condition and cleanliness of food contact surfaces, including utensils, gloves, and outer garments; prevention of cross contamination from insanitary objects to food, food packaging material, and other food contact surfaces, including utensils, gloves, and outer garments, and from raw product to processed product; maintenance of hand washing, hand sanitizing, and toilet facilities; protection of food, food packaging material, and food contact surfaces from adulteration with lubricants, fuel, pesticides, cleaning compounds, sanitizing agents, condensate, and other chemical, physical, and biological contaminants; proper labeling, storage, and use of toxic compounds; control employee health condition and so prevention from the microbiological contamination of food, food packaging materials, and food contact surfaces; removal of pests from the food plant.

2.4.3.2 Monitoring and Records

The processor shall monitor the conditions and practices during processing with sufficient frequency to ensure, at a minimum, conformance with those conditions. Each processor shall correct, in a timely manner, those conditions and practices that are not met. Each processor shall keep SSOP records that, at least the document the monitoring and corrections prescribed.

2.4.4 Hazard Analysis and Critical Control Point

In many countries worldwide, legislation on the safety and suitability of foodstuffs requires hazard analysis and critical control point (HACCP) to be put in place by any food businesses or organizations, whether profit-making or not, whether public or private, carrying out any or all of the following activities: preparation, processing, manufacturing, packaging, storage, transportation, distribution, handling or offering for sale or supply of foodstuffs. According to EU Directive 93/43/EEC on Food Hygiene, all food business operators in the European Union shall implement HACCP. They shall ensure that adequate safety procedures are identified, documented, maintained and reviewed on the basis of the principles used to develop the system of HACCP.

The 7 principles and the guidelines for the application of HACCP have been combined in this specification with basic elements of quality management systems (ISO 9000) to establish "The Requirements for a HACCP based Food Safety System". The 7 principles for HACCP include: conducting hazard analysis and preventive measures; identifying critical control points; establish critical limits; CCP monitoring; corrective actions; record-keeping procedures; verification procedures.

2.4.4.1 Hazard Identification

The food business operator (HACCP team) shall identify and register all potential (biological, chemical and physical) hazards that can have an adverse effect on the safety of the products. The identification shall include all aspects of the operations within the scope of the HACCP system.

The operations to be evaluated include all products, all processes and the prerequisite program of the legal owner of the products. For service organizations (not legal owner, but holder of the products), the hazard identification and analysis is restricted to the services provided, for instance, cold/frozen storage, packaging and transport.

2.4.4.2 HACCP Analysis (Risk)

The food business operator (HACCP team) shall conduct a HACCP analysis to identify which hazards are of such a nature that their eliminations or reductions and control can be kept at acceptable levels, which is essential to the production of safe food.

In conducting the HACCP analysis, the following shall be included:
- the likely occurrence of hazards and severity of their adverse health effects;
- the qualitative and quantitative evaluation of the presence of hazards;
- the concern of the survival or multiplication of micro-organisms;
- the production or persistence of toxins, chemicals or physical agents in foods;
- the conditions that lead to the above.

2.4.5 International Organization for Standardization

The International Organization for Standardization (ISO) is an international standard-setting body composed of representatives from various national standards organizations. Founded on the 23rd of February 1947, the organization promotes worldwide proprietary, industrial and commercial standards. It is headquartered in Geneva, Switzerland.

ISO is an independent, non-governmental organization, the members of which are the standards organizations of the 162 member countries. It is the world's largest developer of voluntary international standards and facilitates world trade by providing common standards between nations. Over twenty thousand standards have been set, covering everything from manufactured products and technology to food safety, agriculture and healthcare.

2.4.5.1 Benefits from ISO Standards for Food

All participants in the food supply chain, whether farmers, manufacturers or retailers, can benefit from the guidelines and best practice contained in ISO standards, which range from food harvesting to product packaging. In addition, internationally agreed standards help food producers meet legal and regulatory requirements.

ISO standards address issues relevant to consumers such as food safety, nutritional labeling, hygiene, food additives and more. They give consumers the peace of mind that comes with knowing the food they consume meets high standards for safety and quality and contains what it says on the lable.

2.4.5.2 ISO Standards for Food

Of the more than 21,500 international standards and related documents, ISO has more than 1,600 standards related to the food sector, with many more in development. These cover: food products, food safety management, microbiology, fisheries and aquaculture, essential oils, starch and its by-products. Many of ISO's standards related to food are developed by ISO technical committee. With its numerous subcommittees and working groups, its responsible scope covers everything from food products, such as coffee, meat, milk, spices and cocoa, to vitamins, animal welfare, microbiology and more.

The ISO 9000 family addresses various aspects of quality management and contains some of ISO's best known standards. The standards provide guidance and tools for companies and organizations who want to ensure that their products and services consistently meet customer's requirements, and that quality is consistently improved.

The ISO 22000 family of International Standards addresses food safety management. ISO 22000 is a certifiable standard that sets out the overall requirements for a food safety management system. It defines the steps an organization must take to demonstrate its ability to control food safety hazards and ensure that food is safe for human consumption. ISO 22000 is one of ISO's best-known standards. Within its broad scope, the ISO 22000 family includes standards specific to catering, food manufacturing, farming, packaging, and animal foodstuffs and feed production.

2.5 Quality Improvement (QI)

Quality Improvement (QI) is a systematic approach that involves mapping, documenting, analyzing, and redesigning. Quality improvement is a revolutionary idea in food quality to raise the level of food quality-no matter how good it may already be-through a continuous search for improvement. In QI, managers, providers, and other staff members are asked not just to meet the standards but rather to exceed them, indeed, to raise the norms.

There are many ways to improve quality, such as enforcing or revising standards, strengthening supervision, and asking managers or technical experts to redesign a process. The concept of QI, which is grounded in the quality movement in industry, usually involves a team-based problem-solving approach.

The QI groups of staff members at the national, district, or facility level work together to identify and resolve problems that affect the quality of foods. They base their decision-making on data rather than on assumptions; they use diagnostic and analytic tools; and they follow a systematic process. An individual supervisor or manager can take this same approach, but QI seeks to harness the managers and personnel at every level to improve the quality of services.

🔊 Lesson 8

Vocabulary

/ Words /

contaminate [kənˈtæmɪneɪt] vt. &vi. 污染
implementation [ˌɪmpləmɛnˈteɪʃən] n. 贯彻，成就
agronomic [ˌægrəˈnɒmɪk] adj. 农艺学的
efflux [ˈɛflʌks] n. 流出
drainage [ˈdreɪnɪdʒ] n. 排水
edaphic [ɪˈdæfɪk] adj. 土壤的

agrochemical [ˌægroʊˈkemɪkl] n. 农用化学品
herbicide [ˈɜːrbɪsaɪd] n. 除草剂
sanitary [ˈsænəteri] adj. 卫生的；清洁的
adulterate [əˈdʌltəreɪt] vt. （尤指食物）掺假

/ *Phrases* /

manure and compost 粪肥和堆肥
nutritional labeling 营养标签
pasture management 牧场管理
conservation tillage 水土保持耕作
surface waters 地表水
organic matter 有机质
soil salinization 土壤盐化
water tables 地下水位
planting time 播种期
stress resistance 抗逆性
crop rotation 作物轮作
hazard analysis 危害分析
critical control point 关键控制点
adverse effect 反作用

/ *Abbreviations* /

good agricultural practice (GAP) 良好农业规范
Food and Drug Administration (FDA) 美国食品与药物管理局
good manufacturing practices (GMP) 良好生产规范
hazard analysis and critical control point (HACCP) 危害分析关键控制点计划
International Food Supplier Standard (IFS) 国际食品供应商标准
safety quality food (SQF) 食品安全与质量保证体系标准
sanitation standard operating procedures (SSOP) 卫生标准操作程序
International Organization for Standardization (ISO) 国际标准化组织

Exercises

I. Write true or false for each of the following statements based on what you have just read.

1. ( ) The aim of quality management is to ensure that products and services are fit for their purpose and meet specifications.

2. ( ) Top decision-makers should have direct experience with day-to-day produce-and-service delivery.

3. ( ) "Customs first" is the first principle in food quality management.

4. ( ) Quality improvement (QI) satisfies regulatory requirements and improves food quality safety.

5. ( ) GMP is the common name given to the sanitation procedures in food production plants.

6. ( ) Good practices related to water will include those that manage ground and soil water by proper use, or avoidance of drainage where required.

7. ( ) ISO is the world's largest developer of voluntary international standards and facilitates world trade by providing common standards between nations.

8. ( ) Chemical, microbial, or extraneous-material testing procedures shall be used where necessary to identify sanitation failures or possible food contamination.

9. ( ) The 7 principles for HACCP including: conduct hazard analysis and preventive measures; identify critical control point; establish critical limits; CCP monitoring; corrective actions; record-

keeping procedures; control risk.

10. (　) SSOP shall address issues relevant to consumers such as food safety, nutritional labeling, hygiene, food additives and more.

11. (　) The ISO 22000 family of International Standards addresses food safety management.

12. (　) Toxic cleaning compounds, sanitizing agents, and pesticide chemicals shall be identified, held, and stored with food.

II. Answer the following questions based on what you have just read.

1. What are the principles of quality management?
2. What are the requirements for good quality control?
3. What is the definition of GAP?
4. What do an individual SSOP include?
5. What are benefits from ISO standards for food industry?
6. How to identify hazards of the HACCP system?
7. List some ISO International Standards which are related to the food sector.

III. Fill in the blanks according to what you have just read.

1. The functions of quality management are _____, _____, _____, _____, and _____.

2. _____, _____, and _____ are commonly used methods for quality assurance policies.

3. _____ helps companies to improve current business processes, whereas _____ satisfies regulatory requirements and improves food quality safety.

4. GMP compliance covers the minimum _____ and _____ that are a prerequisite foundation to further implement other food safety management initiatives such as _____, ISO, IFS and SQF.

5. Cleaning compounds and sanitizing agents used in cleaning and sanitizing procedures shall be free from _____ and shall be _____ and _____ under the conditions of use.

6. GMP ensures that products are consistently _____ and _____ to quality standards.

7. According to EU Directive 93/43/EEC on Food Hygiene all food business operators in the European Union shall implement _____.

8. ISO Standards define the steps an organization must take to demonstrate its ability to control food _____ and ensure that food is safe for human consumption.

IV. Translate the following sentences into Chinese.

1. Quality policy and strategy (QP&S) ensure that the quality management system (QMS) is included in a company's long-term business strategy and helps a company to take appropriate actions and allocate resources to achieve those goals.

2. Sustained success is more likely to be achieved when the organization manages relationships with all of its interested parties to optimize their impact on its performance.

3. soil biotastress resistance

4. verification procedures

5. hazard analysis

6. critical control point
7. adverse effect

V. Translate the following expressions into English.

1. 食品质量管理是以食品链为导向的管理方法。

2. 目前，全世界已经有许多国家对食品的安全性与适宜性进行立法，要求规定，任何食品企业、食品组织，不论是否赢利，不管是国有的还是私人的，都必须执行 HACCP 法则，并应用在食品制备、加工、生产、包装、贮存、运输、销售的全过程。

3. ISO 22000 是 ISO 著名的标准之一。ISO 22000 体系应用广泛，涵盖了餐饮、食品制造、农业、包装、动物食品和饲料生产等方面的标准。

参考译文

第 8 课　食品质量管理

食品质量管理（FQM）是在食品生产系统中以消费者为驱动、以产品为基础的质量管理。它把食品质量在组织和技术方面的概念整合成一个技术-管理概念，并提出了一个综合的观点：食品质量管理是以食品链为导向的管理方法。质量管理所关注的是控制活动，它的目的是确保产品和服务适合管理者的目的和满足产品规格要求。

食品质量管理是指在质量体系内，通过质量策划、质量控制、质量保证和质量改进等手段，制定质量政策、目标和任务的全面管理职能的所有活动。

质量管理的实施取决于组织的规模、供应商和客户的类型、自动化程度、产品类型、质量保证要求，以及最重要的是最高管理层的承诺。

1　质量管理原则

注重质量的组织依靠同样的基本管理原则来取得成功。表 3 列出了质量管理的 7 项原则。

表 3　　　　　　　　　　　质量管理的 7 项原则

	质量原则	描述
1	客户至上	当一个组织吸引并保持客户和其他相关方的信任时，就会取得持续的成功。了解客户和其他相关方当前和未来的需求有助于本组织的持续成功
2	领导作用	领导者建立组织统一的宗旨和方向，所创造的环境能使员工充分参与实现组织的活动
3	全员参与	组织成功的途径是留住称职的员工，鼓励他们不断提高知识和技能，并赋予他们权力，鼓励他们的参与，承认他们的成就

续表

	质量原则	描述
4	过程方法	质量管理体系由相互关联的过程组成。了解该系统是如何产生结果的，使组织能够优化系统及其性能
5	改进	改进对于一个组织保持目前的业绩水平，对其内部和外部条件的变化作出反应，并创造新的机会是必不可少的
6	基于事实的决策方法	决策过程是一个复杂的过程，往往涉及一些不确定性。事实、证据和数据分析使决策过程具有更大的客观性和信心
7	关系管理	利益相关者影响组织的绩效。当组织管理与其所有利益相关方的关系以其对绩效最优化的影响时，更有可能实现持续的成功

2 质量管理的功能

管理是创造性的解决问题。这种创造性的解决问题是通过质量方针和策略、质量策划、质量控制、质量保证和质量改进（图3）。质量改进帮助企业改进现有的业务流程，而质量保证满足监管要求，提高食品质量安全。每一种食品质量管理功能包括多种方法、工具和技术。

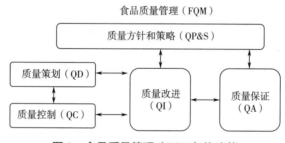

图3 食品质量管理（FQM）的功能

2.1 质量方针和策略

质量方针和策略是确保质量管理体系（QMS）包括在公司的长期业务战略中，并帮助公司采取适当的行动和分配资源以实现这些目标。

质量方针指的是由组织的最高管理者正式发布的该组织总体的质量宗旨和方向。质量管理原则可以作为制定质量方针的依据。对企业来说，质量方针是企业质量行为的指导准则，反映企业最高管理者的质量意识，也反映企业的质量经营目的和质量文化。从一定意义上来说，质量方针就是企业的质量管理理念。

日益激烈的市场竞争和消费者对高质量产品的需求不断增加，迫使食品公司在其战略文件中融入质量价值。全面质量管理（TQM）、质量成本分析（QCA）和战略分析（SA）是企业常用的管理方法。全面质量管理是一个包含组织核心价值、工具和技术的管理系统。质量成本分析的目的是衡量企业在各级运作的有效性，最小化生产成本。同样，公司使用战略分析作为一个过程，建立最合适的管理系统，并以最小的阻力实现业务成果。

2.2 质量策划

质量策划是利用特定的技术手段，将客户的声音转化为产品和工艺的技术要求，使最终的质量达到或超过客户的期望。只有通过确定生产过程的源头，才能保证稳健质量策划达到高质量。

管理人员和计划制定人员通过制定项目中质量的每个方面确保产品的质量和防止问题的产生，包括其方针和目标、资源分配以及标准和指南的制定。

质量策划从定义组织的任务开始，包括它的目的、价值、对象和顾客，并且着眼于质量。这是创造一个新产品和重新设计现有产品的第一步。为了开发与质量有关的现实目标，管理者必须清楚地评定质量水平，并利用现有的资源来达到这个质量水平；机构的优势与劣势，包括当前的程序执行和工作质量；顾客数量，包括顾客本身是怎样认识质量的；以及政治和社会气候。当他们把满足顾客需求作为关注焦点时，自然就会把质量做到最好。

在质量策划时，一线的服务人员、生产人员、现场监管人员以及顾客的参与是至关重要的。最高决策者很少有天天直接参与生产服务的经验。对于从事质量策划的人员来说，没有供应者和顾客的建议，预期的改进是没有意义的，他们的质量策划也不可能吸引顾客或满足人们的需要。

2.3 质量控制

由于食品和生物原料有巨大的差异，质量控制在食品行业中起着重要的作用。质量控制的基本目的是通过采取纠正措施，将变异控制在可容忍的范围内。开发和实施统计和非统计工具和技术来测量、分析和控制食品产品的变化。统计过程控制、验收抽样和目视检验，广泛应用于食品和非食品行业。

质量控制确保项目按计划进行。质量控制活动也可能发现设计上的缺陷，从而指出可以提高质量的变化。对于食品来说，质量控制的主要目标是确保所有过程（从农田到餐桌）都受到控制。质量控制包括日常监督和监测，以确认活动正在按计划进行，工作人员正在遵循指导方针。还包括定期评估，以衡量项目目标的进展情况。良好的质量控制要求项目制定和维护，可衡量的质量指标，及时收集和分析数据，以及有效的监督。

2.4 质量保证

质量保证在食品行业中起着重要的作用，它通过建立标准的组织结构、职责、流程和程序，保证食品安全和可靠性等所有质量义务得到履行。

质量保证是一种在向客户提供解决方案或服务时防止产品错误或缺陷并避免出现问题的方法；ISO 9000 定义为"质量管理的一部分，其重点是提供质量要求得到满足的信心"。质量保证中的缺陷预防与质量控制中的缺陷检测和拒绝有细微区别，并被称为"左移"，因为它在过程的早期关注质量。

质量保证包括两个原则：适合用途（产品应适合预期目的）；正确的第一时间（应消除错误）。质量保证包括对原材料、装配、产品和部件的质量、与生产有关的服务以及管理、生产和检验过程的管理。这两项原则也体现在开发（工程）一种新型技术产品的背景之下：工程的任务是制造。它只起过一次作用，而质量保证的任务是使它始终有效。

为了适应食品行业的发展，目前已经有多个 QA 系统，例如：危害分析关键控制点（HACCP）、国际标准化组织（ISO）、国际食品供应商标准（IFS）、英国零售联盟（BRC）和质量体系（QS）等。

2.4.1 良好农业规范

良好农业规范（GAP）作为一种适用方法和体系，是应用于农场生产，为消费者创造食品或进一步加工食品的一套行为准则。对于由哪些方法组成"良好农业规范"，有着很多不同的定义。生产者可以遵守几个被广泛接受的方案。GAP 指出了减少初级农产品污染可能性的生产方法。实施措施主要涉及到农用水质、粪肥和堆肥的使用、工人健康和卫生以及野生动物、畜禽家畜污染危害控制等。

2.4.2 良好生产规范

良好生产规范（GMP）是确保加工企业生产安全、优质食品的一套操作规范。美国食品与药物管理局（FDA）和全球其他监管机构高度推荐 GMP，以确保产品生产的一致性，满足质量标准。GMP 规范涵盖最低卫生和生产实践要求，是进一步实施其他食品安全管理项目的前提，包括 HACCP、ISO 22000 标准、IFS 和食品安全与质量保证体系标准（SQF）。

2.4.2.1 卫生操作

工厂的建筑物、固定设备及其他有形设施须保持卫生状况，并且保持维修良好，防止食品成为该法案所指的掺杂产品。用具和设备的清洗和消毒须防止对食品、食品接触面或食品包装材料的污染。食品厂内不得存在任何害虫。

用于清洗和消毒的清洗剂和消毒剂不能带有不良微生物，而且须在使用的条件下是安全和合适的。有毒的清洁剂、消毒剂及杀虫剂须被确认、控制和储存，以防止对食品、食品接触面或食品包装材料的污染。

2.4.2.2 加工和控制

食品的进料、检查、运输、分选、预制、加工、包装和贮存等所有操作都须遵守适当的卫生原则。应采用适当的质量管理方法，确保食品适于人们食用，并确保包装材料是安全、适用的。工厂的整体卫生须由一名或数名经指定的、合格的人员进行监督。须采取一切合理的预防措施，确保生产工序不会导致任何来源的污染。必要时，应采用化学的、微生物的或外来杂质的检测方法去确定卫生控制的失误或可能的食品污染。凡是污染已达到该法案所认定的已掺杂的食品都应一律召回，或者，如果允许的话，经过处理或加工以消除该污染。

2.4.3 卫生标准操作程序

卫生标准操作程序（SSOP）是食品生产厂的卫生程序的通用名称。它被认为是 HACCP 的前提方案之一。

SSOP 可以非常简单也可以非常复杂，这取决于它的关注焦点。食品工业设备应采用便于卫生清洁的设计方式，然而，一些有清洗必要的自动化加工设备却很难清洗。一般个性化的 SSOP 应包括：明确需要被清洗的设备或受影响的设备某区域的名称；准备清洗设备或区域所需的工具；如何拆卸设备某区域或整个设备；清洗和消毒的方法。

2.4.3.1 卫生控制

操作程序应包括：确保与食物、与食物表面接触的、被用在冰的生产中水的安全性；保证食品接触面的状态和清洁度，包括器具、手套和外衣清洁；防止因不卫生的物体，如器具、手套和外衣而污染食品、食品包装材料和其他食物接触表面以及生熟制品混放等导致的交叉污染；洗手、手消毒和卫生间设施的维护；保护食品、食品包装材料和食品接触表面不受润滑油、燃料、农药、清洁剂、消毒剂、冷凝液，以及其他的化学、物理和生物污染物的掺入污染；有毒化合物明确标签，安全储存和使用；保障员工的健康，以免引起食品、食品包装材料和食品接触表面的微生物污染；清除食品厂害虫。

2.4.3.2 监测和记录

加工者应频繁利用操作程序，持续监测加工过程的环境和操作，最低限度地确保食品加工在适合的条件下进行。每一个处理者应及时纠正那些不适当的生产环境和操作。加工者应保存SSOP记录，至少包括文档形式的监控和修正记录。

2.4.4 危害分析关键控制点计划

目前，全世界已经有许多国家对食品的安全性与适宜性进行立法，要求规定，任何食品企业、食品组织，不论是否赢利，不管是国有的还是私人的，都必须执行HACCP法则，并应用在食品制备、加工、生产、包装、贮存、运输、销售的全过程。根据欧盟93/43/EEC法规对食品卫生的要求，在欧盟范围内，所有食品经营机构，都必须实施执行HACCP。他们必须确保建立基于HACCP体系法规原则的、有文件记载的、能够识别、维持和评估的适当的安全程序。

HACCP七项原则及应用指南已经包含了建立"基于食品安全的HACCP体系的必备条件"所必须的基础条件——质量管理体系（ISO9000）的基本要素。HACCP七项原则包括：进行危害分析和提出预防措施、确定关键控制点、建立关键界限、关键控制点的监控、纠正措施、记录保持程序、验证程序。

2.4.4.1 危害的识别

企业经营者（HACCP小组）应该能够鉴别并记录所有潜在的对产品安全产生不良影响的（生物的、化学的、物理的）危害。鉴别应该涵盖HACCP体系范围内的所有操作。

评价应包括所有产品、所有过程和法规规定的必备程序。对服务机构（不是合法所有者，但是是产品持有者），危害鉴别和分析仅局限于提供的服务，比如，冷藏/冻藏、包装及运输。

2.4.4.2 HACCP分析（风险评估）

企业经营者（HACCP小组）应该指导HACCP分析以鉴别那些可以消除或者降低及控制在食品安全可接受水平内的危害。

在指导HACCP分析时，应包括以下一些内容：
- 危害的发生及可能，以及对健康产生的不良影响的严重程度；
- 危害存在的定性和定量的评估；
- 对微生物生长和繁殖的关注；
- 产品或食品中毒素、化学或物理试剂的残留；
- 导致以上情况的一些条件。

2.4.5 ISO标准

国际标准化组织（ISO）是由各国标准组织代表组成的国际标准制定机构。成立于1947年

2月23日，该组织促进全球专利、工业和商业标准的发展。它的总部设在瑞士的日内瓦。

国际标准化组织是一个独立的非政府组织，由162个标准化组织成员国组成。它是世界上最大的自愿参与的国际标准化组织，通过提供国家间的共同标准来促进世界贸易。已经制定了超过两万项标准，涵盖从制成品、技术到食品安全、农业和医疗保健的所有领域。

2.4.5.1 ISO标准应用于食品的益处

食品供应链中的所有参与者，无论是农民、制造商还是零售商，都可以受益于ISO标准中包含的准则和最佳操作，这些标准涵盖了食品收获到产品包装等范围。此外，国际商定的标准有助于实现食品加工符合统一的法律和法规要求。

ISO标准可解决与消费者息息相关的问题，如食品安全、营养标签、卫生、食品添加剂等。这种标准化使得消费者安心，因为他们知道他们食用的食品符合安全和质量的高标准，并且包含了标签上所说的所有内容物。

2.4.5.2 食品的ISO标准

在21,500多个国际标准和相关文件中，ISO就占有1,600多个与食品有关的标准，还有更多正在开发中。这些标准包括：食品、食品安全管理、微生物学、渔业和水产养殖、精油、淀粉及其副产品。许多与食品有关的ISO标准都是由ISO技术委员会制定的。它有众多的小组委员会和工作组，负责范围涵盖了从食品（如咖啡、肉类、牛乳、香料、可可粉）到维生素、动物福利、微生物以及更多的领域。

ISO 9000标准体系涉及质量管理的各个方面，包含了ISO一些著名的标准。这些标准为那些希望确保其产品和服务始终如一地满足客户需求的公司和组织提供指导和工具，并不断改进质量。

ISO 22000标准体系讨论食品安全管理。ISO 22000作为一个被认证的标准，规定了食品安全管理体系的总体要求。它定义了一个组织必须采取的步骤，以证明它有能力控制食品可能的安全危害，确保人类消费食物的安全性。ISO 22000是ISO著名的标准之一。ISO 22000体系应用广泛，涵盖了餐饮、食品制造、农业、包装、动物食品和饲料生产等方面的标准。

2.5 质量改进

质量改进是一种系统的方法，它涉及计划、文档化、分析和重新设计。质量改进是食品质量方面的革命性想法。这个想法是通过不断寻求改进来提高食品质量，不管它可能已经有多好的想法。质量改进要求管理人员、供应商和其他工作人员不仅满足标准，而且要超越标准，实际上，要提高标准。

提高质量的方法有很多，例如执行或修改标准，加强监督，并要求管理人员或技术专家重新设计过程。"质量改进"这个概念是以工业中的质量运动为基础的，但通常涉及基于团队的解决问题的方法。

全国、区、工厂等各级的质量改进工作人员共同发现和解决影响食品质量的问题。他们的决策依据是数据而不是假设，使用诊断和分析工具，并遵循系统的流程。个别主管或经理可以采取同样的做法，但质量改进寻求利用各级管理者和个人来提高质量或服务。

LESSON 9

QUALITY CONTROL TOOLS AND FOOD SAFETY EVALUATION

The dictionary defines quality as an important attribute, a degree of excellence or a necessary attribute. A group of activities designed to assure a standard of excellence is called Quality Control. Food is the basic need of life. Quality or excellence in our food supply should be an important concern to all food processors. Safety and wholesomeness are the most important attributes of food quality. The lack of quality as it relates to safety and wholesomeness can result in personal injury, sickness or death. Food-borne illness is an example of sickness or even death when hazardous foods are produced and eaten. Certain foods or food products are defined by regulations or policies which are called standards of identity. These standards of identity are definitions for a specific food product to avoid confusion or mislabeling of similar processed foods.

Another important measure of food quality is government-controlled aspects. Therefore, the first category of food quality is critical attributes and includes factors that affect safety, wholesomeness or legality.

Commitment + Awareness + Teamwork + Communication + Quality Control
= Safe, wholesome and consistent food products

Besides the critical attribute of safety, other properties of the food product should be used to define overall quality. These other features are defined by industry, the processor or consumer demand. Two other categories that classify or describe additional quality characteristics of food products are called major and minor attributes. What is needed for a quality control program? The first step is a strong commitment from management. Quality control must have the same urgency as the profit and loss statement for the business. Quality doesn't cost, it pays. Management must introduce quality awareness throughout the organizational structure, beyond commitment. Finally, to develop a quality control program, it is necessary to define expected food quality to provide a system of quality measurement, allow a means for action not reaction, help to minimize costly errors, and reduce the risk of food safety and wholesomeness defects.

A successful quality control program needs people. It is important that the food operation personnel function as a team and openly communicate to identify problems, issues or opportunities. Once key elements of a quality control program are in place (management commitment, quality awareness, a team effort and open communication), it is able to develop and use additional tools to continue perfect the quality control program.

1 BASIC QUALITY CONTROLS

(1) Ingredient specifications.

(2) Approved supplier list.

(3) Product formulations.

(4) Product standards (specifications).

(5) Manufacturing procedures.

(6) In-process analysis, records and reporting specifications.

(7) Critical control point identification/sampling program.

(8) Packaging and label descriptions.

(9) Good manufacturing practices (GMP) and sanitation.

(10) Warehousing and shipment programs.

(11) Laboratory analysis.

(12) Recall programs.

1.1 Ingredient Specifications

The quality of the finished food products after manufacturing depends on the quality of the raw materials and ingredients. The best starting point for developing ingredient specifications is the supplier. Ask for a copy of the supplier's ingredient specifications. Review the information and modify the specifications to your needs. Discuss and settle specifications with the suppliers. At times, specifications need to be negotiated with suppliers. The ingredient specifications should be documented in a form consistent with the processor's needs. Ingredient specifications documents should include:

①Name of the ingredient.

②Internal code number.

③Expiration date.

④Basic description of ingredient.

⑤Specification categorization: (critical, major, minor).

⑥Action and reject levels.

⑦Ingredient statement.

The prepared ingredient specifications become an effective tool for control of food quality. Under each heading the information should be simple but informative. Critical specifications should include two items associated with public safety.

1.2 Approved Supplier List

For each ingredient, an approved supplier list should exist and be available to individuals responsible for purchasing and quality control. In theory, more than one supplier per ingredient is desirable. A good target is three suppliers per ingredient. A supplier is an ingredient manufacturer, a broker or a distributor. When necessary, identify both the manufacturer and distributor on the approved supplier list. Approve all sources of supply only after careful evaluation and review of their performance in the product. For approving alternate ingredient sources, two key questions are:

Does the ingredient meet the existing or needed specifications?

Does the new ingredient provide the same or desired finished product?

At times, only one acceptable supply source may be available because of special requirements. In this case, alternate sources should be listed for emergent purposes. The emergency source of the ingredient should be one that has been tested and better approaches all specifications.

The approved supplier list should contain the following informations:

①Ingredient name and internal code.

②Supplier name, address, key contact and phone number.

③Trade name of ingredient.

④Supplier code number.

1.3 Product Formulation/Recipe

Trademarked formulas are important. For each food product, written documentation of the formula or recipe should exist and be available for use by selected individuals. The formulas should be used daily as a means to assure consistency between batches, lots and even days of production. Manufacturing personnel need to know the recipe to assure that the product is formulated correctly. Product formulas should include:

①Name of the product.

②Internal code number.

③Expiration date.

④Listing of the ingredients.

⑤Listing of the ingredient codes.

⑥Percentage formula.

⑦Batch formula.

⑧Batch yield.

⑨Batch yield ingredient statement.

Additional information that can be part of a formula document is packaging, lot size, regulatory constraints, net weight, package count per batch, etc.

1.4 Product Standards

A key tool to assure quality in a finished processed food is the product standard document. Product standards define the food by physical, chemical and microbiological characteristics. Appearance, aroma, flavor and texture should also be considered for product standards.

Physical characteristics include size, shape, dimensions, weight, volume, count per package or container, presence of fines, or any other special features which define the particular food. Moisture, fat, protein, ash, fiber and carbohydrate are the basic chemical characteristics. Additional chemical criteria such as salt, sodium, cholesterol, etc., are also used to chemically define food products.

Microbiological standards will be dependent upon the specific food item. First consider food poisoning organisms when developing product standards for a quality control program. Food safety is the responsibility of the processor. If the food product supports the growth of a potential food poisoning organism, identifying the particular organism should be critical in standards category rather than a major or minor standard. Some typical food poisoning organisms are *Salmonella*, *Clostridium botulinum*, *Staphylococcus aureus* and *Clostridium perfringens*.

The sensory properties of a food product are key to the consumer acceptance. Flavor, texture, aroma and appearance are criteria that should be defined to assure that the product meets designed expectations. Qualitative measures of sensory properties can be costly due to requirements for sophisticated equipment. Qualitative testing using taste panels, is an alternative to quantitative measurements. Make a sensory evaluation for flavor, odor and texture a part of a quality control program. Establish a rejection level for each product standard along with acceptable methodology.

1.5 Manufacturing Procedures

For each product, document the method of fabrication or processing procedures to ease duplication from lot to lot, shift to shift and day to day. A simple way to approach this is a clear and concise "cookbook" approach. Key steps in the process which can impact upon yield, quality or production efficiency should be highlighted. Examples of key process steps might be "mix for 3 minutes before adding spices" or "cook to a minimum internal temperature of 62.8 degrees C." Several key points to consider when identifying important processing operations are time, temperature, equipment required, order of addition for ingredients and weight.

1.6 In-process Records

It is important to know what is happening with the product and process during manufacturing. In-process record keeping is a way of obtaining the information. Both quality control and production personnel should participate in maintaining a daily manufacturing log. The specific product weight, temperature, size and shape, ingredient usage, product yield, scrap or waste, material balance and rework are examples of measurements made during the manufacturing process. Base the kinds of in-process measurements used in each operation upon what is called Critical Control Points. In-process record keeping can be a manual or automatic operation and in some cases both. Employee participation in record keeping provides an opportunity for pride in workmanship. In-process records are also a means of making adjustments to the product or process and preventing substandard product. Turn in all in-process records to supervisory management for review at the end of a shift or working day. The supervisory review allows an opportunity to identify problem areas and to make changes to prevent reoccurrence. In some food processing operations, like a poultry or red meat facility, these records are available to the on-site inspector.

1.7 Critical Control Point Identification/Sampling Program

Acritical control point is a step in the process or in product formulation where small differences, changes or mistakes can cause the finished product to be a health hazard, illegal or costly to the business. Critical control points are identifiable. Some critical control points are defined by regulations when public health or product identities are of concern. Cooking temperatures, pasteurization times and temperatures or allowable levels of ingredients are processing variables often defined by regulations. Critical control points may be self-adjusted because of desired label statements on the part of the processor. Net weight is one example while nutritional labeling is another. The cost of a product can be increased by simple employee mistakes. In this case, critical control points in processing simply relate to those processing steps that influence yield or inferior products.

1.8 Packaging and Labeling

A quality control program should include packaging and labeling. The appearance of the package and label are the first elements that the consumer is attracted. Two basic packages are typically necessary for food products. The primary package encloses the food and has direct contact with the product. A film, jar, bottle, carton or box are some of the common primary packages. The secondary package is used to assemble multiple packaged food items for shipment. The shipper or secondary package provides protection, reduces handling of each individual bottle or carton and is necessary for efficient movement of goods to the consumer. Poor packaging or labeling can create negative impressions relative to product quality. This is true for both simple and complex packages and labels. It is to a food processor's advantage to develop packaging and label specifications along the same format as ingredient specifications.

1.9 Good Manufacturing Practices (GMP) and Sanitation

Federal regulations define specific procedures to minimize the contamination of food products by people in manufacturing, processing packaging and warehousing facilities. The regulations are called good manufacturing procedures (GMP). GMP is an integral part of quality control. It is the responsibility of food business management and ownership to know, practice, communicate and ensure that GMP are carried out by employees. An overview of GMP is as follows:

①Individuals with diseases cannot work in the areas where food contamination is possible, such as individuals with infected wounds, sores, and boils.

②Food handlers must follow good personal hygiene practices.

③Train employees effectively on hygiene, sanitation, and pest control.

Along with GMP, a cleaning and sanitizing program is essential. Cleaning and sanitizing should address three basic areas:

①Exterior facilities and grounds.

②Internal facilities including floors, walls, ceilings and ventilation system.

③Equipment and all food contact areas.

The cleaning and sanitizing program controls the dirt and debris, as well as prevents the growth and contamination of microorganisms, insects and pests. It also maintains the equipment in good shape.

1.10 Warehousing and Shipment

Warehousing involves three activities (receiving, storage and shipping) that are included in a quality control program. The receiving operation is the foundation for processing finished food products of a designated quality. Improper storage can adversely impact the quality of materials, ingredients and finished product. Storage in an orderly manner under proper conditions of temperature and humidity is essential to quality. Certain supplies or ingredients may require separation, rotate the inventory. If not properly managed, items may ruin in storage areas. Shipping is the final step in which a food business can have direct control on product quality. Ship items on a first-in-first-out basis and use the same guidelines in shipping that you followed in receiving.

1.11 Laboratory Analysis

The establishment of specifications and standards is meaningless without laboratory analysis or an evaluation program. Laboratory analysis is the phase in which a quality control program is implemented after product is produced. A sampling plan, along with an analysis frequency (time schedule defining how often analysis are made), is absolutely necessary. Compile the methods of analysis used in the laboratory in a special working notebook.

Perform all laboratory analysis in a room away from the processing area. At times, a small food plant may not have a separate area. Therefore, there are three ways to obtain laboratory analysis results:

①In house lab.

②Inside independent lab.

③Combination of in house and independent lab.

Appoint a qualified individual to conduct analyses, report the results and manage the job of quality control. Having laboratory tests results recorded and compared to the specifications or standards. Deviations from standards should be communicated so that additional actions can be taken if necessary.

1.12 Recall Program

Product recall has to bring back products from the distribution system. Every food business is vulnerable to potential product recalls. The public image of businesses can be destroyed during a recall if a well-organized plan is not implemented.

Why would a product be recalled? Products are recalled from distribution areas as a result of voluntary action by a business firm or involuntary action due to government action. Food quality is a demand as well as an expectation of consumers. To meet this consumer need, every food business should develop and use an effective quality control program. Failure to meet consumer demands can cause a decline in product sales and profitability.

2 FOOD SAFETY

Food safety is a scientific discipline describing handling, preparation and storage of food in ways that prevent foodborne illness. The occurrence of two or more cases of similar illnesses resulting from the ingestion of a common food is known as a foodborne disease outbreak. This includes a number of routines that should be followed to avoid potential health hazards. In this way food safety often overlaps with food defense to prevent harm to consumers. The tracks within this line of thought are safety between industry and the market and then between the market and the consumer. In considering industry to market practices, food safety considerations include the origins of food including the practices related to food labelling, food hygiene, food additives and pesticide residues, as well as policies on biotechnology and food and guidelines for the management of governmental import and export inspection and certification systems for foods. In considering market to consumer practices, the usual thought is that food must to be safe in the market and the focus is safe delivery and preparation of the food for the consumer.

The terms of food safety and food quality can sometimes be confusing. Food safety refers to all those hazards, whether chronic or acute, that may make food injurious to the health of the consumer. It is not negotiable. Quality includes all other attributes that influence a product's value to the consumer. This includes negative attributes such as spoilage, contamination by filth, discoloration, off-odors and positive attributes such as the origin, color, flavor, texture and processing methods of the food. This distinction between safety and quality has implications for public policy and influences the nature and content of the food control system which is most suited to meet predetermined national objectives.

Food can transmit pathogens which can result in the illness or death of the person or other animals. Bacteria, viruses, molds and fungus are the main media of transmitting different diseases. It can also serve as a growth and reproductive medium for pathogens.

Another main issue is simply the availability of adequate safe water, which is usually a critical item in the spreading of diseases. In theory, food poisoning is 100% preventable. However, this cannot be achieved due to the number of persons involved in the supply chain, as well as the fact that pathogens can be introduced into foods no matter how many precautions have been taken.

According to WHO, the five key principles of food hygiene are as follows:

①Prevent the contamination of food by pathogens spreading from people, pets, and pests.

②Separate raw and cooked foods to prevent the contamination of the cooked foods.

③Cook foods for the appropriate length of time and at the appropriate temperature to kill pathogens.

④Store foods at the proper temperature.

⑤Usage of safe water and safe raw materials.

Traditional foods are considered safe and nutritious for humans and even though they may contain natural toxins and/or antinutrients. The presumed safety and nutritional value of traditional foods are primarily based on extensive experience and long history of use under well-known conditions of preparation and storage. Therefore, systematic testing of traditional foods is regularly not done unless specific potential risks of consumption are identified, as in the case of foods irradiated for preservation purposes.

Worldwide intensification of agricultural production has put great pressure on basic quality and safety characteristics of food products. Wide-scale use of fertilizers, pesticides, growth promoting agents, and veterinary drugs may leave residues in edible products, and the presence of environmental contaminants in foods may pose possible health risks to consumers and husbandry animals. In addition, pathogenic microorganisms, viruses, and mutant proteins have caused serious outbreaks of human illnesses.

Continuing attention is needed to safeguard the agricultural production chain from current and newly appearing health threats. This is a demanding challenge, given the complexity of agricultural production systems and globalization of trade in agricultural produce.

The development of new food production and breeding technologies using recombinant DNA/RNA technologies also demands our attention, given the fundamentally new characteristics of derived foods and the still very limited experience in evaluating their potential impact on human health and the environment.

Food materials are routinely sampled for a wide range of objectives, which can be broadly categorized into few main areas: risk analysis, compliance with regulatory requirements, post-market monitoring, surveillance, and manufacturing process control. Those involved with safety assessment establish daily requirements for nutrients in food as well as safety ranges for toxic substances that may be present in foods. Those involved with regulatory enforcement, compliance, and post-market monitoring are a diverse and large group that ensures fulfillment of legal requirements at the many different levels of the food production chains.

A reliable analysis of potential health risks for humans consuming foods can only be made when realistic exposure scenarios and well-defined estimations of exposure levels are in place. Various types of substances that are present in the different food matrixes and commodities, raw or semi processed, pose challenges to develop appropriate sampling strategies and analytical detection methods. Both sampling and analytical errors affect the reliability of any final risk estimation, but much more attention has been paid to the development and improvement of analytical methods over the last decades, as compared to the development of appropriate sampling plans. For the last 10~15 years the universal principles of the Theory of Sampling (TOS) have been proven thoroughly, demonstrating that all sampling processes irrespective of the nature of their target lots need to be structurally correct in order to ensure a sufficient degree of accuracy and unbiased, representative precision. This also applies to the assessment of food safety, including food contaminants, additives, naturally occurring toxins/antinutrients, or contaminative microorganisms, and whole foods derived from genetically modified plants.

Risk analysis is a systematic approach to reach conclusions on the safety and nutritional value of foods and associated hazards for humans or animals. It consists of three key fundamentals: "risk assessment, risk management, and risk communication". In the context of risk analysis, a hazard is defined as "the intrinsic potential of a food or agent to cause adverse health effects in humans or the environment". Risk is defined as "the likelihood that under particular conditions of exposure, a hazard will represent a real threat to human or animal health". Risk is thus a risk of hazard and exposure. Risk assessment is a science-driven process, comprised of identification of hazards or potential risks; toxicological/nutritional characterization of the identified hazards; evaluations of exposure to food or associated substances by humans and overall characterization of the identified risks regarding their impact on humans and the environment. The main objective of risk assessment is to characterize the nature and severity of the identified risk and to provide information whether safe threshold levels for consumers can be established, and if these levels have been exceeded.

The risk analysis model, developed by FAO and WHO, is originally designed for the safety assessment of chemical compounds, but is also considered suitable for the assessment of potential food safety risks, which can be of very different nature and origin. The approach is internationally well accepted and used for safety assessment of food related issues. In the improved model, a special evaluation phase is proposed where all available information from the risk-benefit assessment process is evaluated, i. e., acceptability and distribution of risks, costs, and benefits. All interested stakeholders are invited to participate in this transparent evaluation phase to identify possible differences in views on the results of the scientific assessment, which should be considered when final decisions are made by the responsible risk managers.

The improved FAO/WHO risk analysis approach is sufficiently flexible to deal with questions regarding safety of foods contaminated with residues of chemical nature, microbes, or other living materials (viruses and prions), or fortified with nutrients. Foods derived from genetically modified plants or food producing animals, containing new substances or with altered compositions, can also be assessed regarding their safety and nutritional value for humans.

3 FOOD SAFETY HAZARDS

Food hazards are any biological, chemical or physical agent that has the potential to cause an injury or illness if no control mechanism is available. They are the major contributors to food poisoning and could endanger the welfare of the business as well as the quality of different food groups. Food safety hazards are divided into three categories: chemical, physical and biological.

3.1 Chemical Hazards

Chemical contamination occurs as a result of food coming into contact with chemicals. It includes pesticides, food additives, cleaning supplies, and toxic metals that leach from cookware and equipment. Chemical contamination can come from solvents, detergents and sanitizers. Chemicals can leak, seep or give off toxic residues and potentially contaminate food. Chemicals and cleaning equipment need to be stored separately from food and service areas. They also need to be properly labeled. As there is a risk of contamination from pesticides and fertilizers used on raw fruits and vegetables, thus a food handler needs to ensure these items are washed before use.

3.2 Physical Hazards

A physical hazard is an item physically seen in the food. It may enter the food at any stage of production but is most commonly found in the manufacturing or preparation stage. Physical hazards include items: such as hair, fingernails, metal staples, and broken glass, as well as naturally occurring objects, such as bones in fillets.

3.3 Biological Hazards

Undoubtedly, biological hazards pose the greatest threat to food safety till now. Disease-causing microorganisms are responsible for the majority of foodborne illness outbreaks. Biological hazards include certain bacteria, viruses, parasites, and fungi, as well as certain plants, mushrooms, and fish that carry harmful toxins. Food contaminated by bacteria may look completely normal. It cannot be tasted, smelt or seen. However, the consequences will be felt.

4 FOOD TOXICOLOGY

Food toxicology is another term related to food safety. It can be defined as a "systematic study of toxicants found in foods". These compounds can be of natural origin as products of the metabolic processes of animals, plants, and microorganisms from which the food is derived; as biological and chemical contaminants from the air, water, and soil; as intentionally introduced food additives; and as those formed during the processing of foods. Food toxicology is thus concerned with the toxic potential of food, the conditions and factors affecting the presence of these toxicants in food, their interactions with essential dietary nutrients, and the response of the human body to these toxins, and the means of prevention or minimization of these toxic effects as they pertain to food safety and human nutrition.

As compared to the presence of toxicants that are naturally present in various foods, biological contamination of our food supply presents grave food safety concerns. Foodborne diseases caused by bacteria and viruses have varying degrees of severity ranging from mild indisposition to chronic or life-threatening illness. Their importance as a vital public health problem is often overlooked because the true incidence is difficult to evaluate and the severity of the health and economic consequences is often not fully appreciated.

In addition to the foodborne diseases, food contamination caused by mycotoxins, pesticide residues, drug residues, and industrial chemicals is serious issues that affect human safety and well-being. It should, however, be noted that such contamination occurs on a sporadic basis. Furthermore, it can easily be prevented by using careful food production, storage, handling, and preparation practices.

Poison or toxicant is a chemical that is harmful to living organisms because of its detrimental effects on tissues, organs, or biological processes. Any chemical may be a poison at a given dose and a route of administration. Three factors primarily influence the toxicity of any chemical to a given species: the toxic substance itself and the matrix in which it is present, the circumstances of exposure, and the organism and its environment. In practical situations, therefore, the critical factor is not the intrinsic toxicity of a chemical, but rather the risk or hazard associated with its use. In food science and nutrition, it is especially important to understand the concepts of relative risks and safety, hazard, and toxicity associated with the consumption of foods. Risk is the probability that a substance will produce harm under specified conditions. In contrast, absolute safety, or the assurance that damage or injury from the use of a substance, is impossible. However, complete safety is virtually unattainable. Hence, the concept of relative safety has been proposed.

5 EVALUATION

Evaluation means judgment, appraisal, estimation or assessment. It is a process that is used to understand a situation in order to make decisions on whether there is a need to respond to a hazard or to a situation that can lead to a disaster if nothing is done. The assessment must collect information that will allow a good analysis of the situation and the threats to life, human dignity, health and livelihoods of the population. The principle of an assessment is that concerned communities and local authorities are consulted. Through an evaluation it can be determined, in consultation with the relevant authorities and communities, whether assistance is required and, if so, what kind of assistance is needed. In the past, food prepared and used in traditional ways have been judged safe on the basis of long-term experience, even when they contain natural toxicants (e. g. solanine in potatoes) or anti-nutritional substances (e. g. phytate in soya beans). In today's scientific language we would say that a food is considered safe when we are reasonably certain that it will cause no harm if it is used as we intended, under the expected conditions of consumption.

6 HACCP AS A TOOL FOR FOOD SAFETY EVALUATION

HACCP is a management system in which food safety is addressed through the analysis and control of biological, chemical, and physical hazards from raw material production, procurement and handling, to manufacturing, distribution and consumption of the finished product. There are seven discrete actions that are necessary to establish, implement and maintain an HACCP plan, and these are referred to as the "seven principles" in the Codex Guideline (1997). These seven principles are as follows:

Principle 1 - Conduct a Hazard Analysis

The application of this principle involves listing the steps in the process and identifying where significant hazards may occur. The HACCP team will focus on hazards that can be prevented, eliminated or controlled by the HACCP plan. A justification for including or excluding the hazard is reported and possible control measures are identified.

Principle 2 - Identify the Critical Control Points

A critical control point (CCP) is a step at which control can be applied and a food safety hazard can be prevented, eliminated or reduced to an acceptable level. The HACCP team will use a CCP decision tree to help identify the critical control points in the process. A critical control point may control more than one food safety hazard or in some cases more than one CCP is needed to control a single hazard. The number CCP needed depends on the processing steps and the control needed to assure food safety.

Principle 3 - Establish Critical Limits

A critical limit (CL) is the maximum and/or minimum value to which a biological, chemical, or physical parameter must be controlled at a CCP to prevent, eliminate, or reduce to an acceptable level the occurrence of a food safety hazard. The critical limit is usually a measure such as time, temperature, water activity (Aw), pH, weight, or some other measure that is based on scientific literature or regulatory standards.

Principle 4 - Monitor CCP

The HACCP team will describe monitoring procedures for the measurement of the critical limit at each critical control point. Monitoring procedures should describe how the measurement will be taken, when the measurement is taken, who is responsible for the measurement and how frequently the measurement is taken during production.

Principle 5 - Establish Corrective Action

Corrective actions are the procedures that are followed when a deviation occurs in a critical limit. The HACCP team will identify the steps that will be taken to prevent potentially hazardous food from entering the food chain and the steps that are needed to correct the process. This usually includes identification of the problems and the steps taken to assure that the problems will not reoccur.

Principle 6 - Verification

Those activities, other than monitoring, determine the validity of the HACCP plan and that the system is operating according to the plan. The HACCP team may identify activities such as auditing of CCP, record review, prior shipment review, instrument calibration and product testing as part of the verification activities.

Principle 7 - Record Keeping

A key feature of the HACCP plan is recording information that can be used to prove that the food was produced safely. The records also need to include information about the HACCP plan. Record should include information of the HACCP Team, product description, flow diagrams, the hazard analysis, the CCP's identified critical limits, monitoring system, corrective actions, record keeping procedures, and verification procedures.

7 THE IMPORTANCE OF FOOD SAFETY EVALUATION

In order to ensure that consumer food safety education programs are effective in achieving their goals and preventing foodborne illness, it is important that these programs incorporate a rigorous and systematic program evaluation.

Evaluating your education program is important and beneficial to its overall success, for example:

①An evaluation can help you identify the strengths and weaknesses of your program, learn from mistakes, and allow you to continuously refine and improve program strategies.

②By using evaluation data to improve program practices, you can ensure that resources are expended as efficiently and effectively as possible.

③Evaluation data can provide program staff with valuable insights to help them understand the impact of the program, the audience they are serving, and the role they can play to contribute to the program's success.

④There is no way to really know what kinds of effects your program or activities will have without a program evaluation.

⑤Evaluation data can show you how successful your program is in preventing foodborne illnesses and promoting safe food handling practices.

⑥Conducting an evaluation can help you monitor the program and ensure accountability.

🔊 Lesson 9

Vocabulary

/ Words /

specification [spesɪfɪ'keɪʃn] n. 规格
cholesterol [kə'lestərɔːl] n. 胆固醇
Pasteurization [ˌpæstʃərə'zeɪʃn] n. 巴氏杀菌法
contamination [kənˌtæmɪ'neɪʃn] n. 污染
segregation [ˌsegrɪ'geɪʃn] n. 隔离
communicate [kə'mjuːnɪkeɪt] v. 传达
spoilage ['spɔɪlɪdʒ] n. 腐败
discoloration [ˌdɪsˌkʌlə'reɪʃn] n. 褪色

pathogen ['pæθədʒen] n. 病原体
bacteria [bæk'tɪəriə] n. 细菌
fungus ['fʌŋgəs] n. 真菌
preventable [prɪ'ventəbl] adj. 可预防的
evaluation [ɪˌvælju'eɪʃn] n. 评估
solanine [sɒ'lənɪn] n. 龙葵素
phytate ['faɪteɪt] n. 肌醇六磷酸
accountability [əˌkaʊntə'bɪləti] n. 责任制
ownership [oʊnərʃɪp] n. 所有权

/ Phrases /

quality control 质量管理
quality control program 质量控制程序
first-in-first-out（FIFO）先进先出

federal regulations 联邦法规
communicable diseases 传染病
food-borne disease outbreak 食源性疾病暴发

LESSON 9   QUALITY CONTROL TOOLS AND FOOD SAFETY EVALUATION   119

natural toxins 天然毒素
worldwide intensification of agricultural production 全球农业生产集约化
CCP decision tree   CCP 决策树
Codex Guideline 食典指南
pesticide residues 杀虫剂残留
drug residues 药物残留
food safety hazards 食品安全危害
food additives 食品添加剂
pathogenic microorganisms 致病微生物

mutant protein 突变蛋白
derived foods 衍生食品
safety assessment 安全评估
risk assessment 风险评估
risk management 风险管理
risk communication 风险交流
potential food safety risks 潜在食品安全风险
food toxicology 食品毒理学
absolute safety 绝对安全
relative safety 相对安全

/ Abbreviations /

World Health Organization （WHO）世界卫生组织
theory of sampling（TOS）取样理论
Food and Agriculture Organization（FAO）联合国粮农组织

genetically modified（GM）转基因
critical control point（CCP）关键控制点
critical limit（CL）临界极限

Exercises

I. Write true or false for the following statements.

1. (　) Toxic agents can be classified in terms of physical state, effects and their sources.
2. (　) Raw foods and cooked foods should be kept separate.
3. (　) You can tell if food is still safe to eat by smelling it.
4. (　) One of the best ways to prevent contaminating foods is to wash your hands.
5. (　) Foods should be put away in the fridge or freezer within two hours.

II. Multiple choice questions.

1. The most important reason to wash, rinse and sanitize cutting boards is to:

A. Eliminate odors and tastes from getting into other foods

B. Make the cutting boards look better and lasts longer

C. Prevent contamination from one food to another

D. Prevent flavors and garlic or onion juices from onto the others

2. Food handlers should wash their hands after which of the following:

A. Coughing, sneezing, scratching, wiping nose

B. Touching exposed body parts, handling raw animal food

C. Before putting on disposable gloves or after using the rest room

D. All of the above

3. Example of cross contamination is:

A. Raw chicken is processed on a cutting board then lettuce is sliced on same surface

B. Food worker handles raw meat then assembles sandwich without washing hands

C. Liquid from raw hamburger drip onto vegetables for salad

D. All of the above

4. The undesirable change in a food that make it unsafe for human consumption is referred as:

A. Food decay

B. Food spoilage

C. Food loss

D. All of the above

5. Which of the following plays a major role for the food industry?

A. GMP compliance

B. Product recall system

C. Customer service

D. All of the above

6. The model for food safety standard is based on a system called:

A. HACCP –Hazard Analysis and Critical Control Points

B. GMP–Good Manufacturing Practices

C. PHP–Public Health Plan

D. Scientific Studies and Government Regulation

7. Which of these is required on the food label?

A. Total carbohydrate

B. Protein content

C. Fats

D. All of the above

8. To avoid food spoiling in the refrigerator or freezer, which of the following action is always recommended?

A. Placing the oldest product on top of newer product

B. Ordering foods for a few days at a time only

C. Placing a date on arrival of all foods

D. Smelling the food to make sure it is fresh

9. The three categories of food contaminants are best described as:

A. Chemical, biological, and physical hazards

B. Human, microbial, and viral hazards

C. External, internal, and customer hazards

D. Chemical, customer, and staff hazards

10. Aflatoxin is produced by:

A. *Aspergillus* spp.

B. *Salmonella* spp.

C. *Fusarium* spp.

D. *Streptococcal* spp.

Ⅲ. **Answer the following questions shortly.**

1. What is quality control?

2. What is food safety?

3. Mention the basic principles of HACCP.

4. What is toxicity?

5. Define food borne disease.

Ⅳ. **Translate the following words and expressions into Chinese.**

1. Food safety is a scientific discipline describing handling, preparation and storage of food in ways that prevent foodborne illness. The occurrence of two or more cases of a similar illnesses resulting from the ingestion of a common food is known as a foodborne disease outbreak. Considering market to consumer practices, the usual thought is that food ought to be safe in the market and the concern is safe delivery and preparation of the food for the consumer.

2. To reduce the exposure to pesticide, buy and eat organic certified foods as much as possible. If you have a farmer's market in your area, ask them about their use of pesticides, and many smaller farmers choose not to go through the organic certification process, but still do not use pesticides. Always thoroughly rinse or wash all foods before consuming regardless of where it was purchased or grown. Planting your own garden is also another way to be sure you have pesticide-free foods.

Ⅴ. **Translate the following expressions into English.**

1. HACCP 确保食品在消费、生产、加工、制造、准备和食用等过程中的安全，在危害识别、评价和控制方面是一种科学、合理和系统的方法。识别食品生产过程中可能发生风险的主要环节并采取适当的控制措施防止危害的发生。通过对加工过程的每一个步骤进行监视和控制，从而降低危害发生的概率。

2. 食品安全危害主要包括潜在破坏或危及食品安全和质量的因素，包括生物、化学以及物理性的危害，一旦食品含有这些危害因素或者受到这些危害因素的污染，就会成为具有潜在危害的食品。

参考译文

第 9 课　质量控制方法和食品安全评估

　　质量被定义为非常重要的特征，是表示卓越或必要性的属性。旨在确保良好标准的一系列活动称为质量控制。食物是生活的基本需要。食品供应的质量成为所有食品加工商关心的重要问题。安全和健康是食品质量最重要的属性。与食品安全和卫生有关的质量缺乏可能导致人身伤害、疾病或死亡。食源性疾病是食用不安全食物导致疾病甚至死亡的例子。某些食品是由一些法规或政策标准定义的。这些定义的目的是针对一种特定的食品，避免产生误解或对类似产品进行错误标签。

政府控制措施是保障食品质量的另一个重要指标。因此，第一类食品质量是关键属性，包括影响食品安全性的因素、健康或合法性因素等。

承诺+意识+团队合作+沟通+质量控制 = 安全、健康和持续的食品

除了安全的关键属性之外，还应该使用食品的其他属性来定义整体质量。这些属性由行业、加工者或消费者需求来定义。分类或描述食品附加质量特征的另外两类属性称为主要和次要属性。质量控制体系需要什么？首先是管理层的坚定承诺。质量控制必须与企业的盈亏报告一样重要。质量本身虽不花钱，但却会让你付出代价。除承诺外，管理层必须在整个组织机构中灌输质量意识。最后，要完善质量控制体系，必须确定预期的食品质量，并提供相应的质量评估系统及出现问题后的补救措施，从而最大限度减少错误的出现，并降低食品安全和危害健康的风险。

一个完善的质量控制体系需要人员的参与。食品加工部门全体人员作为一个团队发挥作用并集思广益发现问题、矛盾或机遇是非常重要的。一旦质量控制的关键要素到位（管理承诺、质量意识、团队努力和开放沟通），即可开发和使用其他工具继续完善质量控制体系。

1　质量控制的基本要素

（1）成分规格。
（2）批准供应商名单。
（3）产品配方。
（4）产品标准（规格）。
（5）制造工艺。
（6）过程中分析、记录和报告规范。
（7）关键控制点识别/采样程序。
（8）包装和标签规格。
（9）良好生产规范（GMP）和卫生。
（10）仓储和运输。
（11）实验室分析。
（12）召回计划。

1.1　成分规格

食品的质量取决于原材料和成分的质量。制定原料规格的最佳起点是供应商。向供应商索取原料规格的副本。查看信息并根据需要修改规格。与供应商讨论并确定产品规格。有时，规格不仅需要与供应商协商，还应符合加工者需求。原料规格文件应包括：

①配料名称。
②内部代码号码。
③有效日期。
④基本成分描述。

⑤规范分类(重要、主要、次要)。
⑥行动和拒绝水平。
⑦成分说明。

配料规格是食品质量控制的有效工具。每个标题下的信息应该简单但内容丰富。关键规格包括与公共安全相关的两个项目。

1.2 批准供应商名单

对于每种成分,应有一份经批准的供应商名单,供负责采购和质量控制的个人使用。理论上,每种成分不止一个供应商。一个好的目标是每种成分三个供应商。供应商是配料制造商、经销商或分销商。必要时,在批准的供应商名单上标识制造商和分销商。只有在仔细评估和审查产品的性能后,才能批准所有的供应来源。对于批准替代原料来源,两个关键问题是:

成分是否符合现有或需要的规格?

新配料是否提供相同或所需的成品?

有时由于特殊要求,可能只有一个合适的供应来源。在这种情况下,应该列出其他来源用于紧急目的。该成分的应急来源应该经过测试,并且最好接近所有规格检测。

批准的供应商清单应包含以下信息:
①成分名称和内部代码。
②供应商名称、地址、主要联系人和电话号码。
③成分的贸易名称。
④供应商代码编号。

1.3 产品配方

专有的配方是很重要的。每种食品都应有规范化配方的书面文件,并可供特定人员使用。应每日使用该配方以在大量甚至连日的生产中确保批次间的一致性。加工人员需要了解配方以确保产品配制正确。产品配方应包括:
①产品名称。
②内部代码。
③有效期。
④配料表。
⑤配料代码。
⑥配方组成百分比。
⑦批量配方。
⑧批量产量。
⑨批产量成分声明。

配方文件还可包含一定的附加信息,包括包装信息、批量大小、监管约束、净重、每批包装计数等。

1.4 产品标准

产品标准文件是确保加工食品成品质量的关键工具。产品标准通过物理、化学和微生物特征对食物进行规定。食品的外观、香气、风味和质构也可以并且应该是标准考虑的范围。

物理特性包括大小、形状、尺寸、质量、体积、包装大小、是否含有粉末或其他一些特殊食物具有的典型特征。基本的化学指标包括水分、脂肪、蛋白质、灰分、膳食纤维和碳水化合物，另外，盐、钠、胆固醇等化学指标也用于对食品进行化学定义。

微生物指标取决于具体的食品情况。在制定用于质量控制的产品标准时，应首先考虑会引起食物中毒的生物因素。食品安全是加工者应承担的责任，如果食品能够为潜在的有毒微生物提供生长条件，那么对特定生物体的识别应该作为标准类别中的关键部分，而不是主要或次要的标准。一些典型的食物中毒生物体有沙门氏菌、肉毒梭状芽孢杆菌、金黄色葡萄球菌和产气荚膜梭菌。

食品的感官特性是消费者接受的关键，应该确定风味、质地、香气和外观标准以确保产品符合设计预期。由于对先进设备的要求，对感官特性的定性测量可能费用昂贵，可以使用品尝小组进行定性测试代替定量测定。确保对味道、气味和质地进行的感官评估作为质量控制计划的一部分。建立每个产品标准的拒收级别以及可接受的方法。

1.5 制造工艺

记录每个产品制作和加工程序的方法，以帮助在不同批次、班次和每日间重复生产。一个解决这个问题的简单方法是一个清晰和简洁的"食谱"方法。该过程中的关键步骤可以影响产量、质量或生产效率。一个关键工艺步骤的例子是，"加入香料前混合 3min"或"煮至最低内部温度为 62.8℃"。在识别重要的加工操作时，要考虑的几个关键点是时间、温度、所需的设备、配料的顺序和质量。

1.6 进程记录

了解制造过程中产品和工艺的情况很重要。进程中记录的保存是获取信息的一种方式。质量控制和生产人员都应参与日常生产日志的维护。生产过程中进行测量的例子包括具体产品质量、温度、尺寸和形状、配料使用情况、产品产量、废料或废物、材料平衡和返工。根据关键控制点，将每种操作中使用的进程内测量结果作为基础数据。进行中的记录保存可以手动或自动操作，也可两种结合共同进行。员工参与记录提供了一个以工艺为荣的机会。在进行中记录也是对产品或工艺进行调整和防止不合格产品的手段。在轮班或工作日结束时，将所有进行中的记录交给监督管理部门审核。监督审查有机会发现问题并作出改变，以防止再次发生。在一些食品加工业务中，如家禽或红肉设施，这些记录可供现场检查员使用。

1.7 关键控制点识别/采样程序

关键控制点是指产品生产过程中的一个步骤，因为微小的差异、变化或错误可能会导致成品对健康构成危害，对企业而言会造成违法行为并为此付出高额的代价。关键控制点是可识别的。当涉及公共卫生或产品标识时，一些关键控制点是由法律法规规定的。烹调温度、巴氏杀菌时间和温度或原料最大添加量等参数，通常是由法律法规规定的。关键控制点可根据生产者对标签的需要进行自身的调整，例如净重或者营养标签。一个产品的成本可因为员工的小错误而增加。在这种情况下，加工中的关键控制点只与那些影响产量或劣质产品的加工步骤有关。

1.8　包装和标签规格

质量控制应包括包装和标签。影响消费者的第一项内容是包装和标签的外观。食品通常需要两个基本包装。主包装直接与产品接触。薄膜、罐子、瓶子、纸盒或盒子是一些常见的初级包装。二次包装用于组装多个包装食品以供装运。托运人或二级包装提供保护，减少对每个瓶子或纸箱的处理，并且是将货物有效地运送给消费者。相对于产品质量，较差的包装或标签可能会产生负面印象。简单和复杂的包装和标签都是如此。食品加工商的优势在于按照与原料规格相同的格式制定包装和标签规格。

1.9　良好生产规范（GMP）和卫生

联邦法规规定了具体的程序，以尽量减少制造、加工包装和仓储设施中的食品污染。这些规定被称为良好生产程序（GMP）。GMP 是质量控制的组成部分。食品企业管理和所有权人负责了解、实践、沟通并确保 GMP 由员工执行。GMP 概述如下：

①患有传染性疾病的人不能在可能污染食物的区域工作。例如，有创伤、疮、烫伤的人。
②食品加工者必须遵守良好的个人卫生习惯。
③有效地培训员工卫生意识和病虫害防治方法。

参考 GMP，清洁和消毒程序是必不可少的。清洁和消毒应该涉及三个基本领域：

①外部设施和场地。
②内部设施包括地板、墙壁、天花板和通风系统。
③设备和所有食品接触区域。

清洁和消毒程序可防止污垢和碎屑的积聚，防止微生物滋生和污染，并防止昆虫和其他害虫进入和窝藏，维护设备的良好状态。

1.10　仓储和运输

仓储涉及质量控制程序中的三项（接收、贮存和运输）。接收操作是处理指定质量的成品食品的基础。不适当的贮存会对材料、配料和成品的质量产生不利影响。在适当的温度和湿度条件下有序存放对产品质量至关重要。某些用品或配料可能需要隔离、旋转库存。如果没有妥善管理的物品可能会在仓库区域毁坏。运输是食品企业可以直接控制产品质量的最后一步。以先入先出的方式寄出物品，并在收货时遵循相同的装运指南。

1.11　实验室分析

若没有实验室分析或评估程序，规范和标准的制定就没有意义。实验室分析是在产品生产后执行质量控制程序的阶段。一个抽样计划以及分析频率（定义分析频繁程度的时间表）是完全必要的。在特殊工作记录本中编写实验室中采用的分析方法。

在远离加工区域的房间内进行所有的实验室分析。有时，一家小型食品厂可能并没有单独的一个区域。因此，有三种方法可以获得实验室分析结果：

①在室内实验室。
②在独立实验室内。
③室内实验室和独立实验室的结合。

指定一名合格的人员进行分析，报告结果并管理质量控制工作。记录实验室测试结果并与规范或标准进行比较。对测试结果与标准的偏差应该进行有效沟通，以便在必要时采取额外行动。

1.12 召回计划

产品的召回过程预示着将不得不从产品分销体系中带回产品。每个食品企业都受到潜在的产品召回的影响。如果一个组织良好的召回计划没有实施，企业的公众形象可能会在产品召回过程中遭到破坏。

为什么要召回产品？产品从分销中被召回是商业公司的自愿行为或由于政府举措而采取的非自愿行为。高质量食品是消费者所期望的。为了满足消费者的需求，每个食品企业都应该制定和使用有效的质量控制计划。若不满足消费者需求，可能会导致产品销售和盈利能力下降。

2 食品安全

食品安全是描述处理、制备和贮存食品以预防食源性疾病的科学准则。由于摄入相同的食物而导致产生两例及以上相似症状被称为食源性疾病暴发。这包含应遵循的一些准则以避免潜在的健康危害。从这方面来讲，食品安全往往与食品防护重叠，以防止对消费者造成伤害。这条思路中遵循着工厂与市场之间的安全以及市场与消费者之间的安全。在考虑从工厂到市场的行为时，食品安全要考虑的因素包括食品的来源（包括与食品标签、食品卫生、食品添加剂和农药残留有关的做法）、生物技术和食品政策、政府进出口检验管理指南和食品认证系统。在考虑市场对消费者行为时，通常认为食品应该在市场上是安全的，值得关注的是为消费者安全运送和制备食物这一过程。

食品安全和食品质量的概念有时可能会令人困惑。食品安全是指所有那些可能使食物对消费者健康有害的危害因子，无论是慢性的还是急性的。这是不可让步的。质量包括影响产品对消费者价值的所有其他属性。这包括负面属性，例如变质、污染（污物、变色、异味）和积极属性，例如食物的来源、颜色、风味、质地和加工过程。这种食品安全和食品质量之间的区别会对公共政策产生影响，并影响最适合达到预定国家目标的食品控制系统的性质和内容。

食物可以传播能导致人或其他动物生病或死亡的病原体。细菌、病毒、霉菌和真菌是传播疾病的主要媒介，也可以作为为病原体生长和繁殖的基质。

另一个主要问题是充足、安全用水的提供，这通常是疾病传播的关键。理论上来说，可以100%预防食物中毒。然而，由于参与供应链的人数以及无论采取多少预防措施病原体都有可能引入食品，这种理论无法实现。

据世界卫生组织规定，食品卫生的五个关键原则如下：

①防止人类、宠物和害虫传播的病原体污染食物。
②生、熟食物分开，以防止污染熟食。
③通过在适温将食物加热一定的时间来杀死病原体。
④将食物存放在合适的温度下。
⑤使用安全的水和原材料。

人们认为传统食品安全和营养，即使它们可能含有天然毒素和/或抗营养素。基于大众认知的食品加工和贮藏条件下的长期广泛的历史积累经验，来预测传统食品的安全性和营养价值。因此，通常不会对传统食品进行系统检测，除非确定食品存在潜在风险，例如为了贮藏而辐照的食品。

农业生产全球化对食品的质量和安全造成了很大的压力。肥料、农药、生长促进剂和兽药的广泛使用可能会残留在食品中，残留的污染物可能会对消费者和饲养的禽、畜造成健康风险。另外，致病微生物、病毒和突变蛋白（朊病毒）已经引起人类疾病的严重暴发，需要继续关注，保护农业生产链免受当前及新出现的污染物对健康带来的威胁。考虑到农业生产的复杂性和农产品贸易的全球化，这将是一项艰巨的任务。

考虑到新开发食品的基本特点以及评估其对人类健康和环境污染的潜在影响方面的经验仍然非常有限，因此，也需要关注使用重组 DNA/RNA 技术开发食品生产和育种技术。

通常，原材料会被抽样用于各种不同的检测项目，这些项目大致分为几个主要领域：风险分析、产品监管分析、市场监督分析、制造过程控制分析。评估执法人员制定食品中营养素的日常需求及其可能存在有害物质的安全限量范围。这涉及监管执法、合规人员、市场监督人员，可确保食品生产链在各个方面满足法律规范。

只有在实际的暴露情景和对暴露水平有确切估计的情况下，才能对人类食用食物的潜在健康风险进行可靠的分析。存在于不同食品基质和初级或半加工商品中的各种物质对开发适当的取样策略和分析检测方法构成挑战。抽样和分析误差都会影响最终风险评估的可靠性，但与制定适当的抽样计划相比，过去几十年来更多关注分析方法的发展和改进。在过去 10~15 年，抽样理论（TOS）的普遍性原则已被彻底证明，表明所有抽样过程在结构上正确能确保足够的准确度、代表性和无误差时，该过程与其目标批次的性质无关。这也适用于评估食品安全性，包括食品污染物、添加剂、天然毒素/抗营养素或污染微生物，以及源自转基因植物的天然食物。

风险分析是一种系统的方法，可以对人类或动物食品的安全性和营养价值及相关危害作出结论。这包括三个关键要素：风险评估、风险管理和风险沟通。在风险分析的背景下，危害被定义为"食物或制剂对人类或环境造成不良健康影响的内在潜力"。风险被定义为"在特定暴露条件下，危害将代表对人类或动物健康的真正威胁的可能性"。因此风险由危害和暴露产生。风险评估是一个科学驱动的过程，包括识别危害或潜在风险；所识别危害的毒理学/营养鉴定；评估人类接触食物或相关物质的情况，以及所识别风险对人类和环境影响的总体特征。风险评估的主要目标是描述已识别风险的性质和严重程度，并提供信息是否可以建立消费者的安全阈值水平，以及是否已经超过这些水平。

由联合国粮农组织和世界卫生组织制定的风险分析模型最初是为化学化合物的安全评估而设计的，但也被认为适用于评估潜在的食品安全风险，这些风险可能具有非常不同的性质和来源。该方法在国际上被广泛接受，并用于食品相关问题的安全评估。在改进模型中，提出了评估风险-效益评估过程中所有可用信息的特殊评估阶段，即风险、成本和收益的可接受性和分配。邀请所有感兴趣的利益相关者参与这个透明的评估阶段，以确定科学评估结果可能存在的差异，当负责任的风险管理人做出最终决定时应考虑这些差异。

改进的粮农组织/世卫组织风险分析方法具有足够的灵活性，可处理有关化学性质残留物、微生物或其他生物物质（病毒和朊病毒）污染或强化营养素的食品安全问题。即使是含有新物质或改变成分的从转基因植物或食物生产动物衍生的食物，也可以评估其对人类的安全性和营养价值。

3 食品安全危害

食品危害是指如果没有控制机制存在下，任何可能导致伤害或疾病的生物、化学或物理因素。它们是造成食物中毒的主要原因，并可能危及企业的利益以及不同类食物的质量。食品安全危害可分为三类：化学、物理和生物危害。

3.1 化学危害

化学污染是由于食品与化学品接触而发生的。化学危害包括农药、食品添加剂、清洁用品以及从炊具和设备中浸出的有毒金属。化学污染可能来自溶剂、清洁剂和消毒剂。化学品可能泄漏、渗出或放出有毒残留物，并可能污染食物。化学品和清洁设备需要分别存放在食品和服务区。我们还需要正确地对其进行标记。由于水果和蔬菜中使用的农药和肥料有污染风险，因此食品处理人员需要确保这些物品在使用前已被清洗。

3.2 物理危害

物理危害是可以在食物中肉眼观察到的杂质。这些杂质可以在生产的任何阶段被掺杂进入食品，但最常见的是在制造或准备阶段。物理危害包括以下几项，如头发、指甲、金属钉和碎玻璃，以及自然物，例如肉片中的骨头。

3.3 生物危害

到目前为止，生物危害对食品安全构成了最大的威胁。微生物是导致大部分食源性疾病暴发的主要原因。生物危害包括某些细菌、病毒、寄生虫和真菌，以及携带有害毒素的某些植物、蘑菇和鱼类。被细菌污染的食物可能看起来完全正常。但是，我们不能品尝，闻甚至肉眼观察这些食物。否则，将会产生严重后果。

4 食品毒理学

食品毒理学是与食品安全有关的另一个术语。它可以被定义为对食物中发现的有毒物质的系统研究。这些化合物可以是天然来源的，作为动物、植物和食物来源的微生物的代谢过程的产物；可以是来自空气、水和土壤的生物和化学污染物；也可能是有意添加的食品添加剂；以及在食品加工过程中形成的一些产物。食品毒理学由于涉及食品安全与人类营养，因此它与食物的潜在毒性、影响食物中毒物存在的条件和因素、基本膳食营养素的相互作用以及人体对这些毒素的反应、以及预防或最小化这些毒性效应的手段有关。

与各种食物中天然存在的毒物相比，我们食物供应中的生物污染呈现严重的食品安全问题。由细菌和病毒引起的食源性疾病严重程度不同，从轻微不适到慢性或危及生命的疾病。它们作为重要的公共健康问题的重要性常常被忽视,因为真正的发病率难以评估,健康和经济后果的严重程度往往得不到充分的重视。

除了食源性疾病之外,食品中带有真菌毒素、农药残留、药物残留和工业化学品污染也是影响人类安全和健康的严重问题。但是,应该指出,这种污染是不定时发生的。此外,在食物生产、贮存、处理和前期的实践中遵循谨慎的原则可以很容易地避免这种情况的发生。

毒物是对生物体有害的化学物质,可危害生命体的组织、器官或生化系统。在特定剂量和使用途径下,任何化学品都可能成为毒物。化学品对某一物种的毒性主要由以下三个因素决定:有毒物质本身及其基质、有毒物质暴露的环境、有机体及其所处环境。因此,在实际情况中,相对化学品的内在毒素,更应关注化学品使用过程可能产生的风险或危害。在食品科学和营养学中,了解与食品消费相关的相对风险、安全性、危害以及毒性的概念尤为重要。风险是物质在特定条件下会产生危害的概率。相反,绝对安全是保证使用某种物质不会造成任何伤害。然而,绝对安全实际上是难以达到的。因此,提出相对安全的概念。

5 评估

评估意味着判断、评价、预估或评定。这是一个通常用于了解情况的过程,以便决定是否需要对危害或者可能导致危害的环境作出反应。评估必须收集信息,以便对威胁生命、人类尊严、健康和人口生计的危害因素进行良好分析。评估的原则是咨询受影响的社区和地方相关部门,通过评估,可以与相关部门和社区协商确定是否需要援助,如果需要,什么样的援助最合适。过去,食品的加工与食用安全都是依据长期经验为基础来判断评估,甚至食品本身有天然毒物(例如马铃薯中的龙葵素)或抗营养物质(例如大豆中的植酸盐)。在今天的科学语言中,我们认为只有食品在我们预期消费条件下无毒无害,才说明他们是安全的。

6 食品安全评价工具——HACCP

HACCP是一种食品安全管理体系,从原材料生产、采购及处理到成品的生产、分销及销售过程,每一个方面都对生物、化学及物理危害进行分析和控制,以确保食品的安全。制定、实施和保持HACCP计划需要7种措施,在食典指南(1997)中被称为"七项原则",具体如下:

原则1 进行危害分析

这一原则的应用包括列出加工过程的步骤,并确定可能会发生显著危害的地方。HACCP小组将重点关注能够被预防、消除和控制的危害,对于危害物的判断理由需要记录,并确定合理的控制措施。

原则2 确定关键控制点

关键控制点(CCP)指可以应用控制措施的点、步骤或程序,或者可以将食品安全危害预防、消除或减少至可接受的水平。HACCP小组将使用CCP决策树来帮助识别加工过程中的关键控制点。关键控制点可以控制多个食品安全危害,或者在某些情况下,需要多个CCP来控制单一危害。CCP 需要的数量取决于加工过程和确保食品安全所需的控制。

原则 3　建立关键限值

关键限值（CL）是 CCP 必须控制的生物、化学或物理参数的最大值及/或最小值，以预防、消除或减少食品安全危害发生的可接受水平。关键限值是一项衡量指标，如时间、温度、水分活度（Aw）、pH、质量或基于科学文献、监管标准的其他一些指标。

原则 4　关键控制点监控

HACCP 小组将描述在每个关键控制点测量关键限值的监测程序。监测程序应描述如何进行测量，何时进行测量，谁负责测量以及在生产过程中测量的频率。

原则 5　建立纠正措施

纠正措施是在超出关键限值时进行的程序。HACCP 小组定义其为防止潜在危害食品进入食物链将采取的步骤以及纠正过程所需的步骤。这通常包括明确问题并采取措施确保问题不再发生。

原则 6　验证

这些活动（监控除外）决定 HACCP 计划的有效性，并且系统按照计划运行。HACCP 小组可以认定一些活动，如对 CCP 的审核、记录审核、装运前审查、仪器校准和产品测试等，作为验证活动的一部分。

原则 7　保持记录

HACCP 计划的一个关键组成部分是记录可用于证明食品安全生产的信息。记录还需要包含有关 HACCP 计划的信息。记录应包括 HACCP 小组、产品描述、流程图、危害分析、CCP 确定的关键限值、监测系统、纠正措施、记录保存程序和验证程序等信息。

7　食品安全评估的重要性

为了确保消费者食品安全教育计划能够有效实现其目标并预防食源性疾病，将这些计划纳入严格和彻底的计划评估是极其重要的。

评估教育计划对其整体成功非常重要且有益。例如：

①评估有助于确定计划的长处和短处，从错误中吸取教训，从而不断完善和改进计划策略。

②通过使用评估数据来改进计划的实施，可以尽可能有效和高效地使用资源。

③评估数据可以为项目工作人员提供宝贵的见解，帮助他们了解项目的影响、他们所服务的受众以及他们可以发挥的作用，为项目的成功做出贡献。

④没有方案评估，无法真正了解计划或活动会产生什么样的影响。

⑤评估数据可以显示计划在预防食源性疾病和促进安全食品处理方面的成功程度。

⑥进行评估可以监控计划并确保其可追溯性。

LESSON 10

CRISIS MANAGEMENT FOR FOOD INDUSTRY

1 CRISIS DEFINITIONS

A crisis is defined by the dictionary as a "critical moment or turning point". A business book, on the other hand, might define a crisis as a substantial, unforeseen circumstance that can potentially jeopardize a company's employees, customers, products, services, fiscal situations, or reputation. Both definitions contain an element of urgency that requires immediate decisions and actions from people involved.

The definition given by the American Institute for Crisis Management (ICM) for the word "crisis" underscores the association of a crisis with media coverage. ICM defines "crisis" as "a significant business disruption which stimulates extensive news media coverage. The resulting public scrutiny will affect the organization's normal operations and could also have a political, legal, financial, and governmental impact on its business".

The basic causes of a business crisis are four in number:
(1) Natural disaster (storms, earthquakes, volcanic action, etc.)
(2) Mechanical problems (ruptured pipes, metal fatigue, etc.)
(3) Human errors (the wrong valve was opened, miscommunication about what to do, etc.)
(4) Management decisions/indecision (the problem is not serious, nobody will find out)

Most of the crises ICM have studied fall in the last category and are the results of management not taking action when they were informed of a problem that eventually would grow into a crisis. Crisis events generally fall into two basic types based on the length of warning time.

1.1 Sudden Crisis

A sudden crisis is defined as: a disruption in the company's business which occurs without warning and is likely to generate news coverage and may have adverse impact on the following aspects:
(1) Our employees, investors, customers, suppliers or other publics;
(2) Our offices, franchises or other business assets;
(3) Our revenues, net income, stock price, etc.;
(4) Our reputation and the ultimate good will be listed as an asset on our balance sheet.

A sudden crisis may be:

(1) A business-related accident resulting in significant property damage that will disrupt normal business operations;

(2) The death or serious illness or injury of management, employees, contractors, customers, visitors, etc. as the result of a business-related accident;

(3) The sudden death or incapacitation of a key executive;

(4) Discharge of hazardous chemicals or other materials into the environment;

(5) Accidents that cause the disruption of telephone or utility service;

(6) Significant reduction in utilities or vital services needed to conduct business;

(7) Any natural disaster that disrupts operations, endangers employee;

(8) Unexpected job action or labor disruption;

(9) Workplace violence involving employees/family members or customers.

1.2 Smoldering Crisis

A smoldering crisis is defined as: any serious business problem that is not generally known inside or outside the company, which may generate negative new-coverage if or when it goes "public" and could result in more than a predetermined amount in fines, penalties, legal damage awards, unbudgeted expenses and other costs. Examples of the type of smoldering business crises that would prompt a call to the crisis management team would include:

(1) Sting operation by a news organization or government agency;

(2) OSHA or EPA violations which could result in fines or legal action;

(3) Customer allegations of overcharging or other improper conduct;

(4) Investigation by a federal, state or local government agency;

(5) Action by a disgruntled employee such as serious threat or whistle-blowing;

(6) Indications of significant legal/judicial/regulatory action against the business;

(7) Discovery of serious internal problems that will have to be disclosed to employees, investors, customers, vendors and/or government officials.

In some instances crisis situations may be either sudden or smoldering, depending on the amount of notices in advance and the chain of events in the crisis. Examples would include:

Adverse government actions	Computer tampering
Anonymous accusations	Damaging rumor
Competitive misinformation	Discrimination accusations
Confidential information disclosed	Equipment, product or service sabotage
Misuse of chemical products	Industrial espionage
Disgruntled employee threats	Investigative reporter contact
Death or serious injury of employee	Judicial action
Employee involved in a scandal	Labor problem

Licensing disputes with local officials	Lawsuit likely to be publicized
Extortion threat	Security leak or problem
False accusations	Severe weather impact on business
Incorrect installation of equipment	Sexual harassment allegation
Grand jury indictment	Special interest group attack
Grass roots demonstrations	Strike, job action or work stoppage
Illegal actions by an employee	Terrorism threat or action
Indictment of an employee	Illegal or unethical behavior of an employee
Malfunction of major equipment	Union organizing actions
Nearby neighbor, business protest	Whistle-blower threat or actions

Many empirical researches into business crisis events indicate that most sudden crises also generate "aftershocks" in the form of smoldering crises which occur as the government, media and internal investigations into the cause of the crisis and uncover specific problems that were not known previously. Many of those aftershocks are included in the list directly above.

2　CRISIS MANAGEMENT

Crisis management is the process of preparing for and responding to an unpredictable negative event in order to prevent it from escalating into an even bigger problem, or worse, exploding into a full-blown, widespread, life-threatening disaster. Crisis management involves the execution of well-coordinated actions to control the damage and preserve or restore public confidence in the system under crisis.

In the context of corporate governance, excellent crisis management is a "must" whenever a crisis occurs because of its enormous potential impact on the company's reputation and financial standing. Poor handling of a crisis situation can ruin the confidence of the customers or the public in a company and jeopardize its survival, and it normally takes a long time to correct, if it still is repairable at all. This is the importance of public perception of a company's handling of a crisis situation that media coverage management has become an important ingredient of crisis management.

Crisis management doesn't start only when a crisis arises and ends when "the last fire has been put out". Crisis management requires actions before a crisis happens, while the crisis is unfolding, and after the crisis has ended. In fact, crisis management is divided into three stages:

(1) Pre-incident stage, which involves identification of potential crisis situations and developing contingency plans for responding to each of them;

(2) Incident stage, which involves management of an ongoing actual crisis situation itself;

(3) Post-incident stage, which includes corrective and preventive actions to preclude the recurrence of the same crisis situations and business recovery actions to restore public confidence in the brand

or the company.

There are many different ideas or theories on how to best manage a crisis situation. These differing ideas, nonetheless, have some common elements:

(1) The need to anticipate potential crisis situations and prepare for them;

(2) The need to provide accurate information during a crisis;

(3) The need to react as quickly as possible to the situation;

(4) The need for a response that comes from the top;

(5) The need for long-term solutions.

Anticipating potential crisis situations that a company may encounter and formulate as well as document contingency action plans for them are a basic requirement of any crisis management program. These plans should also be well-rehearsed by all employees, so the conduct of regular drills are also needed. Any company must be prepared to deal with fires, bomb threats, personnel violence and natural disasters such as earthquakes and tornadoes. In the semiconductor industry, the discovery of a life-threatening device reliability issue, processing gas leaks and hazardous chemical spills, even the sudden loss of a major supplier, are examples of crisis situations.

One of the hardest things to contend with during a crisis is the craving of customers or the public for a constant supply of information. Accurate information pertaining to a crisis is often not readily available at once, and is difficult to announce to the public once they become available. If the crisis does not concern the public, then a company may stay quiet about it.

Otherwise, the press will usually be all over the place within hours. In such a case, there is no choice—most experts agree that it is better to provide accurate information, no matter how painful they sound, than to manipulate the situation by giving false information to the public, which often backfires with tremendous repercussions. Thus, a company that has developed a culture of internal secrecy and manipulation is of great disadvantage in this respect, because they would find it difficult to provide honest information and subsequently resort to the thing they usually do: hide the truth. Public relation work for high-profile crisis is something that any management team must be well-trained for.

Being indecisive, playing time or inability to get accurate information quickly can be disastrous during a crisis. Management must act swiftly and decisively to contain the problem, assess affected goods, ensure business continuity, allay public fears, and preserve company reputation even while the crisis is still unfolding. Since a crisis by definition is unpredictable, a company needs to have a system for assembling a crisis management team that knows what to do within an hour after a crisis occurs, 24 hours, all day, and all year long.

The visibility of top management during a crisis is highly recommended by experts because it assures the public that the problem is getting due corporate attention. Management must also actively pursue long-term corrective and preventive actions to avoid falling in the same crisis situation again.

A comprehensive crisis management program includes the following components:

(1) An emergency response, which consists of all activities pertaining to safe management of immediate physical, health, and environmental effects of the crisis;

(2) Business continuity, or the company's ability to continue delivering goods and services despite

the crisis;

(3) Crisis communications, which pertains to the internal and external public relations (PR) management activities during a crisis;

(4) Humanitarian assistance, or the company's efforts to alleviate the physical, emotional, and psychological effects of the crisis on other people;

(5) Drills and exercises that allow personnel to rehearse what they need to do in a crisis situation

3　INTERNATIONAL CRISIS MANAGEMENT

International crisis management takes many forms. In the food industry, the primary focus involves product quality issues related to product contamination, mislabeling, and product adulteration, which may result in the withdrawal or recall of products from the market. The movement of food products between countries greatly increases the complexity.

Once a crisis situation is identified, it is critical to have a comprehensive crisis management plan to guide the company through the event. All functional groups in the company will be involved in the investigation, execution, or public relations of a recall. It is essential that each group and individual knows their role and has the resources available to accomplish their required tasks. Outside third-party resources will often be required to support these efforts: public relations, media contacts, legal counsel, government liaisons, testing laboratories, university personnel, retrieval services, and crisis consulting services.

The crisis management plan must be an "evergreen" document, which is continuously maintained and updated to address the changing regulatory environment as well as internal personnel changes. "recall simulations" should be a part of the plan to regularly practice execution of the plan and to train personnel in real life situations. The regulatory requirements and the consumer environment vary greatly in different world areas.

4　CRISIS MANAGEMENT IN FOOD ENTERPRISES

Food industry is a moral industry. The crisis management of food enterprises must be emphasized and put in the leading position of enterprises. The crisis of food enterprises is not equal to the failure of enterprises. Food enterprises are constantly seeking technological innovation, market management and organization system. Meanwhile, crisis management innovation should be put in an important position. Successful businesses can not only handle crises properly, but also turn crises into business opportunities.

Common methods of crisis management in food enterprises:

(1) Quick recovery of unqualified products;

(2) Give material and spiritual compensation to the persons concerned;

(3) Actively use the media to guide the public;

(4) Use authoritative opinions to deal with the crisis;
(5) The use of legal crisis regulation;
(6) Announce the cause of the crisis;
(7) Reshape the good public image.

At present, there are some outstanding problems in the management of food enterprises in China, which need to be standardized and implemented with the awareness, methods and means of crisis management. When the food crisis has occurred, the enterprise should take emergency action in accordance with the pre-established early warning scheme to control or eliminate the crisis that has occurred and reduce the loss caused by the crisis. Although crises are inevitable, they can be managed. A positive crisis response can minimize the losses caused by the crisis.

Suggestions on the countermeasures of food enterprise crisis management:
(1) Safe and qualified food is the cornerstone of enterprise crisis management;
(2) Firmly establish crisis awareness and strengthen safety culture construction;
(3) Improve the level of prediction and decision;
(4) Establish a complete set of measures to deal with crises and rebuild the new image of enterprises.

🔊 Lesson 10

Vocabulary

/ *Words* /

jeopardize [ˈdʒepərdaɪz] vt. 危害
fiscal [ˈfɪskl] adj. 财政的，会计的
reputation [ˌrepjuˈteɪʃn] n. 名誉，名声
score [skɔːr] v. 划线于……，强调；底线
scrutiny [ˈskruːtəni] n. 详细审查
earthquake [ˈɜːrθkweɪk] n. 地震
volcanic [vɑːlˈkænɪk] adj. 火山的
rupture [ˈrʌptʃər] v. 破裂，裂开，断绝（关系等）；割裂
fatigue [fəˈtiːg] v. 疲乏，疲劳，累活

valve [vælv] n. 阀
indecision [ˌɪndɪˈsɪʒn] n. 优柔寡断
asset [ˈæset] n. 资产，有用的东西
contractor [kənˈtræktər] n. 订约人，承包人
endanger [ɪnˈdeɪndʒər] v. 危及
smolder [ˈsmoʊldər] v. 郁积，闷烧
whistleblowing [ˈwɪslbˌloʊɪŋ] n. 告密，揭发
aftershock [ˈæftərʃɑːk] n. 余震

perception [pər'sepʃn] n. 感知，感觉
contingency [kən'tɪndʒənsi] n. 偶然，可能性，意外事故，可能发生的附带事件
rehearse [rɪ'hɜːrs] v. 预演，排演，使排练，复述，练习
tornado [tɔːr'neɪdoʊ] n. 旋风，龙卷风，大雷雨，具有巨大破坏性的人（或事物）
craving ['kreɪvɪŋ] n. 渴望
pertaining [pər'teɪnɪŋ] adj. 与……有关系的，附属……的，为……固有的（to）
manipulate [mə'nɪpjuleɪt] vt. （熟练地）操作，使用（机器等），操纵（人或市价、市场）
backfire [ˌbæk'faɪər] v. 逆火，产生反效果，事与愿违

repercussion [ˌriːpər'kʌʃn] n. 弹回，反响，（光、声等的）反射
swiftly ['swɪftli] adv. 很快地，即刻
allay [ə'leɪ] v. 减轻，减少
unfold [ʌn'foʊld] v. 打开，显露，开展，阐明
visibility [ˌvɪzə'bɪləti] n. 可见度，可见性，显著，明显度，能见度
humanitarian [hjuːˌmænɪ'teriən] n. 人道主义者
liaison ['liːəzɑːn] n. 联络，（语音）连音
retrieval [rɪ'triːvl] v. 取回，恢复，修补，重获，挽救，拯救

/ Phrases /

smoldering crisis 潜在危机   recall simulations 模拟召回

/ Abbreviations /

Occupational Safety and Health Act (OSHA) 职业安全与卫生条例【美】   Environmental Protection Agency (EPA) 美国环保署

Exercises

I. Answer the following questions according to the passage.

1. What is the definition of crisis? How to comprehend its meaning by yourself?
2. What is the definition of sudden crisis? Please give several examples of sudden crisis.
3. What are basic causes of a business crisis?
4. What is the definition of smoldering crisis? Please give several examples of smoldering crisis.
5. What is the definition of crisis management? How to execute crisis management?
6. What are common elements of crisis management?
7. How many components had been included a comprehensive crisis management program?
8. How to implement a crisis management for food industry?

II. Choose a term from what we have learned to fill in each of the following blanks and change the word form where necessary.

1. A crisis can potentially _____ a company's employees, customers, products, services, fiscal

situation, or _____ .

2. Earthquakes are acts of _____ , metal fatigue is a _____ problem and miscommunication is a human error.

3. Crisis events generally fall into two basic types based on the amount of warning time, they are _____ and _____ .

4. Unexpected job action or labor disruption is a _____ crisis, investigation by a federal, state or local government agency is a _____ crisis.

5. Most sudden crises generate " _____ " in the form of smoldering crises.

6. Crisis Management is the process of preparing for and responding to an _____ event.

7. Crisis management is divided into three stages: _____ , _____ and _____ .

8. Within the food industry, the primary focus involves product quality issues related to product contamination, _____ and _____ .

参考译文

第 10 课　食品工业危机管理

1　危机的定义

危机的词典定义为"关键时刻或转折点"。一本商业书籍可能会将危机定义为一种可能危及公司员工、客户、产品、服务、财政状况或声誉的不可预见的重大事件。这两种定义都包含一种紧迫的因素，需要人们立即作出决定和采取行动。

美国危机管理协会（ICM）对"危机"一词的定义，强调了危机与媒体报道的关联。ICM 将"危机"定义为"一场重大的业务中断，刺激了广泛的新闻媒体报道。由此引发的公众监督将影响该组织的正常运作，也可能对其业务产生政治、法律、金融和政府影响。"

商业危机的基本原因有四个：

（1）不可抗力（如风暴、地震、火山活动等）；

（2）机械故障（如管道破裂、金属疲劳等）；

（3）人的错误（如阀门被错误地打开，错误的沟通应该做什么等）；

（4）管理决策优柔寡断（如问题不严重，没有人会发现）。

ICM 所研究的大多数危机都属于最后一类，它们是管理层在被告知一个最终会发展成危机的问题时不采取行动的结果。危机事件通常根据预警时间的多少分为两种基本类型。

1.1　突如其来的危机

一场突如其来的危机被定义为：公司业务的中断，没有预警，可能产生新闻报道，并可能在以下方面产生负面影响：

(1) 员工、投资者、客户、供应商或其他公众；
(2) 办公室、专营权或其他商业资产；
(3) 收入、净收入、股票价格等；
(4) 声誉和最终的好处将被列为资产负债表上的资产。

突然的危机可能是：
(1) 与商业有关的事故，造成重大财产损失，将扰乱正常的经营活动；
(2) 因与业务有关的事故而导致的死亡或严重疾病或管理层、雇员、承包商、顾客、访客等的伤害；
(3) 一名关键行政人员的突然死亡或能力丧失；
(4) 将危险化学品或其他材料排放到环境中；
(5) 造成电话或公用事业服务中断的事故；
(6) 为开展业务所需的公用事业或重要服务的显著减少；
(7) 任何破坏经营、危及员工的自然灾害；
(8) 意外的工作行动或劳动中断；
(9) 工作场所暴力涉及员工/家庭成员或客户。

1.2 潜在危机

潜在危机定义为：任何在公司内部或外部不为人知的严重的业务问题，如果公开，可能产生负面新闻报道，可能导致超过预定金额的罚款、处罚、法律损害赔偿金，预算外的费用和其他费用。那些会引发对危机管理团队的呼吁的潜在业务危机的例子包括：
(1) 由新闻机构或政府机构进行的突击行动；
(2) OSHA 或 EPA 违规行为，可能导致罚款或法律诉讼；
(3) 客户对收费过高或其他不当行为的指控；
(4) 联邦、州或地方政府机构的调查；
(5) 一个心怀不满的雇员的行为，比如严重的威胁或泄密；
(6) 对该业务有重大法律/司法/监管行动的迹象；
(7) 发现严重的内部问题，必须向员工、投资者、客户、供应商和/或政府官员披露。

在某些情况下，危机的情况可能是突然的，也可能是潜在的，这取决于提前通知的数量和危机中的事件链。例子有：

不良政府行为	电脑篡改
匿名指控	损害谣言
竞争的错误信息	歧视指控
机密信息披露	设备，产品或服务破坏
滥用化学产品	工业间谍
不满的员工	威胁调查记者
雇员死亡或严重伤害	司法行为
卷入劳工丑闻	雇员问题

与地方官员的诉讼	纠纷可能会被公开
敲诈勒索的威胁	安全漏洞或问题
诬告	恶劣天气对商业造成严重影响
安装不正确的设备	性骚扰指控
大陪审团起诉	特别利益集团的攻击
基层示威	罢工、临时性罢工、停工
雇员的非法行为	恐怖主义威胁或行动
对雇员的控告	雇员违法或不道德行为
主要设备故障	工会组织的作用
附近的邻居、企业抗议	告密者威胁或行动

许多对商业危机事件的实证研究表明，大多数突发危机也会产生"余震"，以潜在危机的形式存在。在政府、媒体和内部对危机起因的调查中，会发现一些以前不知道的具体问题。这些余震中有许多都直接列在上面的名单中。

2 危机管理

危机管理是对不可预测的负面事件进行准备和应对的过程，以防止它升级为更大的问题，甚至更糟，爆发成一场全面、广泛、危及生命的灾难。危机管理包括采取协调一致的行动来控制损害，保护或恢复公众对危机中系统的信心。

在公司治理的背景下，危机发生时，优秀的危机管理是"必须"的，因为危机对公司的声誉和财务状况产生了巨大的潜在影响。对危机情况的糟糕处理可能会破坏客户或公众对公司的信心，危及公司的生存，通常需要很长一段时间才能纠正，如果它仍然是可以挽回的。这就是公众对公司处理危机情况的认识的重要性，媒体报道管理已经成为危机管理的重要组成部分。

危机管理并不是只有当危机发生时才开始，当"最后的火被扑灭"时才会结束。危机管理需要在危机发生前、危机正在展开时、危机结束后采取行动。事实上，危机管理分为三个阶段：

（1）事件前阶段　包括识别潜在危机情况，制订应对每一种危机情况的应急计划；

（2）事件阶段　涉及对正在进行的实际危机情况的管理；

（3）事件后阶段　包括纠正和预防措施，防止再次发生相同的危机情况和业务恢复行动，以恢复公众对品牌或公司的信心。

关于如何最好地应对危机，有许多不同的想法或理论。然而，这些不同的观点有一些共同的因素：

（1）需要预见潜在的危机情况并为之做好准备；

（2）危机期间提供准确信息的需要；

（3）需要尽快对情况作出反应；

(4) 来自高层的响应需求;
(5) 需要长期的解决方案。

预见公司可能遇到的潜在危机情况,制订并记录应急行动计划是任何危机管理计划的基本要求。这些计划也应该由全体员工精心排练,因此也需要进行定期演练。任何公司都必须准备好应对火灾、炸弹威胁、人员暴力以及地震和龙卷风等自然灾害。在半导体行业,发现危及生命的设备可靠性问题、处理气体泄漏、危险化学品泄漏,甚至是主要供应商的突然损失都是危机情况的例子。

在危机中最难应付的事情之一是客户或公众对信息不断提供的渴望。与危机有关的准确信息往往不容易立即获得,而且一旦它们变得可用,就很难向公众宣布。如果这场危机与公众无关,那么一家公司可能会对此事保持沉默。

否则,媒体通常会在几个小时内就报道。在这种情况下,没有选择的余地——大多数专家都认为,提供准确的信息,无论听起来多么痛苦,都比通过向公众提供虚假信息来操纵局势更好,而这些信息往往会产生巨大的反响。因此,一个开发了一种文化的内部保密和操纵的公司在这方面有缺点,因为他们会发现很难提供真实信息,随后诉诸他们通常做的事:隐藏真相。任何管理团队必须对引人注目的危机公共关系工作训练有素。

在危机中,优柔寡断、拖延时间或无法快速获得准确的信息可能是灾难性的。管理层必须迅速果断地采取行动、控制问题、评估受影响的产品、确保业务连续性、消除公众的恐惧,并在危机尚未爆发时保持公司声誉。由于定义上的危机是不可预测的,公司需要建立一个系统来组建一个危机管理团队,在危机发生后一个小时内、24小时内、全天内、全年内都知道该做什么。

在危机期间,高层管理人员的可见性是专家极力推荐的,因为它向公众保证,问题正得到企业的关注。管理层还必须积极寻求长期的纠正和预防措施,以避免再次陷入同样的危机状况。

一个全面的危机管理计划包括以下部分:
(1) 紧急响应,包括有关安全管理危机对身体、健康和环境的直接影响有关的所有活动;
(2) 业务连续性,或公司在危机期间继续提供商品和服务的能力;
(3) 危机公关,涉及危机期间的内部和外部公关管理活动;
(4) 人道主义援助,或公司为减轻危机对他人身体、情感和心理影响的努力;
(5) 演习和练习,能够让人员演练在危急情况下需要做什么。

3 国际危机管理

国际危机管理有多种形式。在食品行业内,主要关注产品质量问题,涉及产品污染、错贴、产品掺假,这可能导致产品从市场上退出或召回,国家间食品的流动大大增加了管理的复杂性。

一旦确定了危机情况,制订一个全面的危机管理计划来指导公司渡过难关至关重要。公司内的所有职能部门都将参与召回的调查、执行或公共关系。重要的是,每个小组和个人都知道自己的角色,并拥有完成所需任务的资源。通常需要第三方资源来支持你的努力:公共关系、媒体接触、法律顾问、政府联络人、检测实验室、大学人员、检索服务和危机咨询服务。

危机管理计划必须是一份"常青"文件，它将不断地维护和更新，以应对不断变化的监管环境以及内部人事变动。"召回模拟"应该是计划的一部分，定期执行计划并在现实生活中培训人员。在不同的世界范围内，监管要求和消费环境差别很大。

4　食品企业危机管理

食品工业是道德工业，食品企业的危机管理必须得到重视，并且要放在企业的首要位置。食品企业发生危机并不等同于企业失败。食品企业在不断谋求技术、市场管理和组织制度等一系列创新的同时，应将危机管理创新放到重要位置上。成功的企业不仅能够妥善处理危机，而且能够化危机为商机。

食品企业危机管理的常用方法：
（1）迅速召回不合格产品；
（2）给有关人员予以物质和精神补偿；
（3）积极利用传媒引导公众；
（4）利用权威意见处理危机；
（5）利用法律调控危机；
（6）公布造成危机的原因；
（7）重塑良好公众形象。

目前我国食品企业管理中存在一些突出的问题，需要以危机管理的意识、方法和手段加以规范和实施。当已经发生食品危机事件后，企业应迅速根据事先制定的预警方案，采取应急行动，控制或消除已经发生的危机事件，减轻危机带来的损失。危机虽然不可避免，但却是可以管理的，积极的危机应对可将危机造成的损失降到最低。

食品企业危机管理对策的建议：
（1）安全合格的食品是企业危机管理的基石；
（2）牢固树立危机意识，强化安全文化建设；
（3）提高预测、决策水平；
（4）建立一整套危机处理措施，重塑企业新形象。

References

[1] Knechtges, P. L. Food Safety-Theory and Practice [M]. Sudbury: Jones & Bartlett, 2011.

[2] Ian C. S. Food Safety-The Science of Keeping Food Safe [M]. Chichester: Wiley-Blackwell, 2018.

[3] Forsythe S. J. The Microbiology of Safe Food [M]. Chichester: Wiley-Blackwell, 2000.

[4] Young W. P., George F. W. H. Milk and Dairy Products in Human Nutrition [M]. Chichester: Wiley-Blackwell, 2013.

[5] Norman N. P., Joseph H. H. Food Science [M]. 5th ed. Berlin: Springer, 2012.

[6] Alcides T. Safe food-A practical approach to food safety [M]. Saarbrucken: LAP Lambert Academic, 2014.

[7] Marshall, R. J, McElhatton A. Food safety-A practical and case study approach [M]. Berlin: Springer, 2007.

[8] Holban A. M. Food safety and preservation-modern biological approaches to improving consumer health [M]. London: Academic Press, 2018.

[9] Giuseppe M. Anthocyanins in Fruits, Vegetables, and Grains [M]. Oxfordshire: Taloy & Francis Ltd, 2018.

[10] William J. S., Debbie N. Egg Science and Technology [M]. 4th ed. Holland: Haworth Inc, 1995.

[11] Ashurst. P. R. Chemistry and Technology of Soft Drinks and Fruit Juices [M]. Chichester: John Wiley & Sons Inc, 2016.

[12] Sunwoo, H. H. Chemical Composition of Eggs and Egg Products [J]. Handbook of Food Chemistry, 2014: 1-27.

[13] Young W. P. Milk and Dairy Products in Human Nutrition: Production, Composition and Health [M]. Chichester: Wiley-Blackwell, 2013.

[14] Norman N. P., Joseph H. H. Food Science [M]. 5th ed. Berlin: Springer, 2012.

[15] Noble P. W. Fundamentals of Dairy Chemistry [M]. NewYork: Van Nostrand Reihold Co., 1988.

[16] Hui, Y. H. Meat Science and Applications [M]. NewYork: Narcel Dekker, 2001.

[17] Fidel T. Handbook of Meat Processing [M]. Ames: Wiley-Blackwell, 2010.

[18] Park, Y. W. Haenlein. G. F. Handbook of Milk of Non-Bovine Mammals [M]. Ames: Wiley-Blackwell, 2006.

[19] Press, C. Food safety in China: A mapping of problems, governance and research [Z]. Section 3. 2014.

[20] 陈冠林，高永清. 我国食品安全问题频发的原因及对策 [J]. 中国食物与营养, 2012, 18 (3): 5-8.

[21] 中华人民共和国卫生部. 卫生部办公厅关于 2012 年全国食物中毒事件情况的通报

[J]. 中华人民共和国国家卫生和计划生育委员会公报, 2013, 20 (2): 5-6.

[22] Hefnawy, M. Advances in Food Protection: Focus on Food Safety and Defense [M]. 2011.

[23] Miraglia M, Marvin H. J. P, Kleter G. A et al. Toxicity assessment of unintentional exposure to multiple chemicals [J]. Toxicol Appl Pharmacol, 2007, 223: 104-113.

[24] Trichopoulos D. Epidemiology of Cancer. In: DeVita VT (ed) Cancer: principles and practice of Oncology [M]. Lippincott, Philadelphia. 1997.

[25] Turkdogan M. K, Kilicel F, Kara K et al. Heavy metals in soil, vegetables and fruits in the endemic upper gastrointestinal cancer region of Turkey [J]. Environ Toxicol Pharmacol. 2002, 13: 175-179.

[26] Sharma R. K, Agrawal M, Marshall F. M. Heavy metal (Cu, Zn, Cd and Pb) contamination of vegetables in urban India: a case study in Varanasi [J]. Environ Pollut, 2008, 154: 254-263.

[27] Rangan C, Barceloux DG. Medical toxicology of natural substances: foods, fungi, medicinal herbs, toxic plants, and venomous animals. In Food contamination: chemical contamination and additives [J]. Wiley, Hoboken, 2008, 5-21.

[28] Harner T, Pozo K, Gouin T et al. Global pilot study for persistent organic pollutants (POPs) using PUF disk passive air samplers [J]. Environ Pollut, 2006, 144: 445-452.

[29] Tareke E, Rydberg P, Karlsson P et al. Analysis of acrylamide, a carcinogen formed in heated foodstuffs [J]. J Agric Food Chem, 2002, 50: 4998-5006.

[30] Sanny M, Luning P. A, Marcelis WJ et al. Impact of control behaviour on unacceptable variation in acrylamide in French fries [J]. Trends Food Sci Technol, 2010, 21: 256-267.

[31] Tardiff RG, Gargas M. L, Kirman CR et al. Estimation of safe dietary intake levels of acrylamide for humans [J]. Food Chem Toxicol, 2010, 48: 658-667.

[32] Stoker C, Rey F, Rodriguez H et al. Sex reversal effects on Caiman latirostris exposed to environmentally relevant doses of the xenoestrogen bisphenol-A [J]. Gen Comp Endocrinol, 2003, 133 (3): 287-296.

[33] Palanza P, Howdeshell K. L, Parmigiani S et al. Exposure to a low dose of bisphenol A during fetal life or in adulthood alters maternal behaviour in mice [J]. Environ Health Perspect, 2002, 110 (suppl 3): 415-422.

[34] Dessi-Fulgheri F, Porrini S, Farrabollini F. Effects of perinatal exposure to bisphenol A on play behaviour of female and juvenile rats [J]. Environ Health Perspect, 2002, 110 (suppl 3): 403-407.

[35] Lopez-Cervantes J, Paseiro-Losada P. Determination of bisphenol A in, and its migration from PVC stretch film used for food packaging [J]. Food Addit Contam, 2003, 20 (6): 596-606.

[36] Ozaki A, Baba T. Alkylphenol and bisphenol A levels in rubber products [J]. Food Addit Contam, 2003, 20 (1): 92-98.

[37] Brede C, Fjeldal P, Skjevrak I et al. Increased migration levels of bisphenol A from polycarbonate baby bottles after dishwashing, boiling and brushing [J]. Food Addit Contam, 2003, 20 (7): 684-689.

[38] Gossner CM-E, Schlundt J, Embarek PB et al. The melamine incident: implications for international food and feed safety [J]. Environ Health Perspect, 2009, 117 (12): 1803-1808.

[39] Brown C A, Jeong K-S, Poppenga RH et al. Outbreaks of renal failure associated with melamine and cyanuric acid in dogs and cats in 2004 and 2007 [J]. J Vet Diagn Invest, 2007, 19: 525-531.

[40] Smith J, Hong-Shum L (eds). Food additives data book. Blackwell Science, Oxford. 2003.

[41] Bouwmeester H, Dekkers S, Noordam MY et al. Review of health aspects of nanotechnologies in food production [J]. Regul Toxicol Pharmacol, 2009, 53: 52-62.

[42] Lyndhurst B. An evidence review of public attitudes to emerging food technologies [M]. UK: Food Standards Agency. 2009.

[43] Siegrist M. Factors influencing public acceptance of innovative food technologies and products [J]. Trends Food Sci Technol, 2008, 19: 603-608.

[44] Tauxe RV, Doyle MP, Kuchenmüller T et al. Evolving public health approaches to the global challenge of foodborne infections [J]. Intern J Food Microbiol, 2010, 139 (1): 516-528.

[45] Pasqualina Laganà, Emanuela Avventuroso, Giovanni Romano, Maria Eufemia Gioffré, Paolo Patanè, Salvatore Parisi, Umberto Moscato, Santi Delia. Chemistry and Hygiene of Food Additives: Food Preservatives [M]. Italy: Springer, 2017: 5-9.

[46] Wu Jie, Sun Bei, Zhu Fei. The Development and Application Prospect of Natural Food Preservative [C], China: Springer, 2012: 404-412.

[47] 郝利平. 食品添加剂 [M]. 北京: 中国农业大学出版社, 2016 (7): 34-54.

[48] 孙平. 食品添加剂 [M]. 北京: 中国轻工业出版社, 2020 (5): 22-41.

[49] 魏明英, 翟培. 食品添加剂应用技术 [M]. 北京: 科学出版社, 2020 (9): 12-27.

[50] CAC. Codx Alimentarius Commission procedural mannual. 20th ed.: Codex Alimentarius Commission. 2011.

[51] 陈松, 翟琳. 欧盟食品风险评估制度的构成及特点分析 [J]. 农业质量标准. 2008 (05): 54-56.

[52] 王祎, 刘俊荣, 朱蓓薇. 水产食品"风险-收益"分析研究进展 [J]. 水产科学, 2015, 34 (9): 589-596.

[53] 杨杏芳, 吴永宁, 贾旭东. 食品安全风险评估—毒理学原理、方法与应用 [M]. 北京: 化学工业出版社, 2017.

[54] 陈君石, 石阶平. 食品安全风险评估 [M]. 北京: 中国农业大学出版社, 2018.

[55] 吴元元. 信息基础、声誉机制与执法优化—食品安全治理的新视野 [J]. 中国社会科学, 2012 (6): 115-133.

[56] 可山. 食品安全管理研究: 现状述评、关键问题与逻辑框架 [J]. 管理世界, 2012 (10): 176-177.

[57] Schwägele F. Traceability from a European perspective [J]. Meat Science, 2005, 71 (1): 164.

[58] 王成, 赵多勇, 王贤, 等. 食品产地溯源及确证技术研究进展 [J]. 农产品质量与安全, 2012 (S1): 59-61.

[59] 魏益民, 郭波莉, 魏帅, 等. 食品产地溯源及确证技术研究和应用方法探析 [J]. 中国农业科学, 2012, 45 (24): 5073-5081.

[60] 王世鹏, 董文宾, 樊成, 等. 非线性化学指纹图谱技术在食品掺假检测中的应用 [J]. 食品研究与开发, 2016 (1): 204-207.

[61] 朱潘炜, 刘东红, 黄伟, 等. 指纹图谱技术在食品品质检测中的应用 [J]. 粮油加工, 2008 (6): 125-128.

[62] 赵海燕, 郭波莉, 张波, 等. 小麦产地矿物元素指纹溯源技术研究 [J]. 中国农业科学, 2010 (18): 3817-3823.

[63] 王洁, 伊晓云, 马立锋, 等. ICP-MS 和 ICP-AES 在茶叶矿质元素分析及产地溯源中的应用 [J]. 茶叶学报, 2015 (3): 145-150.

[64] 胡桂仙, GOMES A. H, 王俊, 等. 电子鼻无损检测柑橘成熟度的实验研究 [J]. 食品与发酵工业, 2005, 31 (8): 57-60.

[65] 张宁, 张德权, 李淑荣, 等. 近红外光谱结合 SIMCA 法溯源羊肉产地的初步研究 [J]. 农业工程学报, 2008, 24 (12): 309-311.

[66] 张萍, 闫继红, 朱志华, 等. 近红外光谱技术在食品品质鉴别中的应用研究 [J]. 现代科学仪器, 2006 (1): 60-62.

[67] 陈全胜, 赵文杰, 张海东, 等. SIMCA 模式识别方法在近红外光谱识别茶叶中的应用 [J]. 食品科学, 2006, 27 (4): 186-189.

[68] 鹿保鑫, 张东杰. 基于矿物元素指纹图谱的黑龙江黄豆产地溯源 [J]. 农业工程学报, 2017, 33 (21): 216-221.

[69] 李平惠, 钱丽丽, 杨义杰, 等. 基于矿物元素指纹图谱技术的芸豆产地溯源研究 [J]. 中国粮油学报, 2016, 31 (6): 134-139.

[70] 赵雅楠, 王颖, 张东杰. 基于 SSR 标记的黑龙江省绿豆品种遗传多样性分析及指纹图谱构建 [J]. 食品工业科技, 2017, 38 (19): 148-153.

[71] Olaimat, A. N. Holley. R. A. Factors influencing the microbial safety of fresh produce: A review [J]. Food Microbiology, 2012, 32: 1-19.

[72] 许学勤. 食品专业英语文选 [M]. 北京: 中国轻工业出版社, 2014.

[73] 张兰威, 李佳新. 食品科学与工程英语 [M]. 哈尔滨: 哈尔滨工程大学出版社, 2007.

[74] 殷文政, 樊明涛. 食品微生物学 [M]. 北京: 科学出版社, 2015.

[75] 江汉湖. 食品微生物学 [M]. 北京: 中国农业出版社, 2008.

[76] 柳增善. 食品病原微生物 [M]. 北京: 中国轻工业出版社, 2007.

[77] Luning, P. A. Food quality management: A technomanagerial approach [M]. Netherlands: Wageningen Pars, 2002.

[78] Paster, T. Food safety for the 21 stcentry: managing HACCP and food safety throughout the global supply chain [M]. New Jersey: Wiley-blackwell, 2013.

[79] WHO Guidelines on Good Agricultural and Collection Practices (GACP) for Medicinal Plants-2004 [S].

[80] Michael, C. Food Plant Sanitation: Design, Maintenance, and Good Manufacturing Prac-

tices, Second Edition [M]. Florida: CRC Press, 2013.

[81] The International Organization for Standardization (ISO): Global Governance through Voluntary Consensus-2009 [S].

[82] Carol A. Wallace, William H. Sperber, Sara E. Mortimore. Food safety for the 21 st century: Managing HACCP and food safety throughout the global supply chain [M]. Wiley-Blackwell (John Wiley and Sons Ltd.), 2012 (4). 279-285.

[83] RichParker. 江波译. Introduction To Food Science (影印版) [M]. 北京: 中国轻工业出版社 2005 (7): 235-242.

[84] 陈忠军. 食品专业英语 [M]. 北京: 中国林业出版社, 2016 (6): 33-45.

[85] 宁喜斌. 食品安全风险评估 [M]. 北京: 化学工业出版社, 2017 (8): 52-63.

[86] Djidiack Faye. Food safety and quality assurance issues-Improving the Safety and Quality of Fresh Fruits and Vegetables: A Training Manual for Trainer [M]. NewYork: Food Science and Technology, 2007.

[87] National Research Council (NRC). Ensuring Safe Food: From Production to Consumption [R]. Washington: National Academy Press, 1998.

[88] Potter, Morris E. Factors for the Emergence of Foodborne Disease. In Proceedings of the Fourth ASEPT International Conference [J]. Food Safety 1996: 185-195.

[89] Tauxe, R., H. Kruse, C. Hedberg et al. Microbial Hazards and Emerging Issues Associated with Produce. A preliminary report to the National Advisory Committee on Microbiologic Criteria for Foods [J]. Journal of Food Protection 1997. 60 (11): 1400-1408.

[90] Food Quality and Safety Systems—A Training Manual on Food Hygiene and the Hazard Analysis and Critical Control Point (HACCP) System. Food Quality and Standards Service Food and Nutrition Division, Food and Agriculture Organization of the United Nations Rome, 1998.

[91] Foster, E. M. Historical overview of key issues in food safety [J]. Emerg. Infect. 1997. Dis. 3: 481-482.

[92] Thorne, Stuart. The History of Food Presentation. Totowa, NJ: Barnes and Noble Books. 1986.

[93] Mottram, D. S, Wedzicha, B. L, Dodson, A. T. Acrylamide is formed in the Maillardreaction [J]. Nature, 2002, 419: 448-449.

[94] Stadler R. H., Blank, I. Varge, N et al. Acrylamide from Maillard reaction products [J]. Nature, 2002, 419: 449-450.

[95] FAO/WHO. Summary and Conclusions of Joint FAO/WHO committee on food additives sixty-fourth meeting Rome, February 8-17 2005, JECFA/64/SC.

[96] European Food safety Authority, Statements of the scientific panel on contaminants in the food chain a summary report on acrylamide in food of the 64th meeting of the Joint FAO/WHO expert committee on food additives. 2005.

练习题答案

LESSON 1　OVERVIEW OF FOOD SAFETY

I. Write true or false for the following statements according to the passage.

1. T　2. F　3. T　4. T　5. F　6. F　7. T　8. T

II. Answer the following questions according to the passage

1. The most frequent causes of foodborne disease are diarrheal disease agents, particularly *campylobacter* spp. In addition, such as non-typhoidal *salmonella enterica*, are also responsible for the majority of deaths due to foodborne disease in all regions of the world. Another cause of foodborne disease is mycotoxins.

2. Persistent organic pollutants (POPs) are a class of organic compounds that are produced industrially for use as insecticides and as plasticizers in a variety of products. Long-term exposure to POPs has been linked to reproductive disorders, immune system dysfunction, nerve damage and even cancers.

3. The major heavy metals of concern in the food pollution are lead, cadmium, mercury, and arsenic. Additionally, Chromium selenium, tin, antimony, copper, thallium, fluoride, and zinc also pose potential health threats to consumers.

III. Fill in the blanks according to the passage

1. maize, rice, sorghum
2. lead, cadmium, mercury

IV. Translate the following words and expressions into Chinese

食源性疾病	收获后
微生物活性	微生物
成品，制成品	环境污染
持久性有机污染物	有机化合物
食品添加剂	抗生素

V. Translate the following expressions into English

1. Heavy metal emissions from processing facilities enter the environment and contaminate air and soil which can lead to the contamination of drinking water and food crops. In some regions in China, heavy metals have caused serious agricultural land and food pollution.

2. There are many diseases resulting from the consumption of food containing pathogenic micro-organisms or their toxic metabolites. Because microorganisms are abundant in the field and water where foods are produced harvested, they are commonly associated with the finished products.

LESSON 2　FOOD RAW MATERIAL INGREDIENT AND NUTRITION

I. Write true or false for the following statements according to the passage

1. F　2. T　3. F　4. T　5. T　6. F　7. T　8. F　9. F　10. T

Ⅱ. Answer the following questions according to the passage.

1. Illness and death from diseases caused by contaminated food. The most frequent causes of foodborne disease are diarrheal disease agents. Another cause of foodborne disease is mycotoxins.

2. To prevent diseases caused by pathogenic bacteria in milk and to lengthen the shelf life of milk. Because milk samples from normal healthy mammary glands contain many strains of bacteria.

3. Each kernel of grain is composed of three parts: the germ, endosperm, and bran, and if all are present in a grain, it is a "whole grain," such as whole wheat.

4. Cereals lack lysine, the most important and scarce essential amino acid in human nutrition. Therefore, protein quality of cereals is low and it is affected by the digestibility rate and mainly by the essential amino acid balance.

5. Fruits contain mostly sugars and fibers that are extensively fermented in the large intestine, especially apples and pears, are rich in fructose. Free fructose is poorly absorbed and its function is similar to dietary fiber, escaping absorption in the small intestine while being fermented in the large intestine.

Ⅲ. Fill in the blanks according to the passage

1. 3.3 g/100g

2. Collagen, gelatin

3. protein, riboflavin

4. digestible carbohydrates, fiber

5. caseins, whey proteins, non-protein nitrogen

Ⅳ. Translate the following words and expressions into Chinese

动物蛋白　　　　　　　　　　食品添加剂
环境条件　　　　　　　　　　高价值动物蛋白
食源性感染　　　　　　　　　非蛋白氮
重金属　　　　　　　　　　　低脂动物产品
健康威胁　　　　　　　　　　乳清蛋白

Ⅴ. Translate the following expressions into English

1. Food scientists generally agree that there are at least six important attributes of food: safety, purity convenience in use, shelf life, functional performance, and nutritional value. Here we are dealing with a wide range of important characteristics that largely determine the acceptability of food by most consumers.

2. One of the main quality problems of human food supply is the contents of vitamin and mineral. The aim of human food intake from a biological perspective is to survive. Human has adapted to the surrounding environment in the long time of evolution process, and gradually formed nutritional requirements of the model.

LESSON 3　ISSUES IN FOOD SAFETY

Ⅰ. Write true or false for the following statements according to the passage.

1. T　2. T　3. T　4. T　5. T　6. T　7. T　8. F　9. T　10. T　11. F　12. F　13. T　14. T　15. T　16. T　17. F　18. T　19. T　20. T

II. **Answer the following questions according to the passage.**

1. Food safety issues reflected there were many hidden dangers in today's food safety, the main reasons including: inferior raw materials are used in the process of food manufacture, addition of toxic substances and hard to stop, overuse of food additives, abuse of non-food processing chemical additives, the safety situation of agricultural products and poultry products is also not optimistic, antibiotics, hormones and other harmful substances remain in poultry, livestock and aquatic products. Genetically modified foods have the potential threat, though there is not enough evidence to prove that genetically modified foods are harmful to humans.

2. Food safety issues occurred in the entire food supply chain, every links possible occurred food safety issues including materials growth environment, climate and soil, process, package, storage, transportation and sale.

3. Emerging food processing technologies threat will pose by to food safety issues including two parts:

(1) New physical treatments

High pressure pasteurization: incomplete microbial inactivation, incomplete spore inactivation, some chemical reactions.

Pulsed electric fields: incomplete spore inactivation, electrochemical reactions; metal transfer from electrodes.

Cold plasma: free radicals, oxidations.

Light pulses and UV: photo-oxidative reactions.

Ultrasound: microbial inactivation.

Ohmic heating: metal transfer from electrodes.

Ionising radiation: free radicals, oxidative reactions.

(2) New chemical treatments

Modified atmosphere: can delay microbial growth, but tends to over-extend shelf-life and storage time.

Anti-microbial agents: incomplete microbial inactivation; microbial growth only delayed in storage, potential toxicity.

4. Agro-food technologies tend to elicit consumer rejection including:

Ionising irradiation of foods;

Hormonal (and antibiotic) treatment of animals to hasten growth and increase meat or milk production (banned in the EU);

Various food additives;

Excessive use of crop fertilisers and pesticides;

Genetically modified food (GMF) crops and food ingredients;

Genetically modified animals (including cloned animals).

5. Active food package refers to which can supply the information of packaged food properties during the circulation and storage periods by detecting the enviroment conditions of packaged food, such as time-temperature indicator, ripeness indicator, and packaging leakage, etc. Active food package bring

safety risk to food safety issues is that migration of chemicals and their degradation products into the food.

Ⅲ. Fill in the blanks according to the passage.

1. heat stress, irradiation, plant fertilizer, toxic substances, microorganisms, parasites, insects, worms

2. heavy metals, inorganic anion, organic pesticides and polychorobiphenyls

3. acrylamide, bisphenol A

4. carbohydrate-rich foods, potatoes and cereal-based products

5. fertilizers, pesticides, drugs and packaging materials

Ⅳ. Translate the following words and expressions into Chinese.

食品安全事件　　　　　　　　　　转基因食品
非食品加工化学添加剂　　　　　　知情选择权
食物供应链　　　　　　　　　　　食品活性包装
工业废水　　　　　　　　　　　　即食食品
新兴食品技术　　　　　　　　　　上市前授权程序

Ⅴ. Translate the following expressions into English.

1. They also provide scientific advice on many food safety issues such as food additives, chemical and microbiological contaminants, and agro-chemical residues.

2. Food safety issues are related closely to people's daily lives, paying little attention will lead to irreparable disaster.

3. Along with economic development, food consumption is being increasingly diversified and consumers are increasingly on food safety issues, which calls for higher demand of food logistics.

4. The consumer's "right to be informed choice" is well established in the EU. Mandatory labelling for irradiated foods and GM foods has discouraged manufacturers and retailers to place such foods on the market.

LESSON 4　FOOD PRESERVATIVES

Ⅰ. Write true or false for the following statements according to the passage.

1. T　2. F　3. F　4. T　5. T

Ⅱ. Multiple choice questions (Choose the correct answers according to the passage).

1. C, D　2. A, B, C, D　3. B, D

Ⅲ. Answer the following questions according to the passage.

1. Food preservatives are a series of substances added to food to prevent decomposition and prolong the shelf life of the food.

2. Stable property, and effective within a certain period of time;

Non-toxic during usage or after decomposition, and do not hinder the normal function of gastrointestinal enzymes, nor affect the intestinal normal probiotics;

Bacteriostatic or bactericidal effect at low concentrations;

No pungent and abnormal smell;

Easy to use, and have a reasonable price.

3. (1) Understand the physical and chemical properties of food preservatives, such as solubility, heat resistance, optimal pH, antibacterial spectrum and minimum inhibitory concentration.

(2) Understand the quality of food itself and the condition of the bacteria.

(3) Understand the environmental conditions in food processing, storage and transportation, and to ensure the best preservative performance of food preservatives.

4. (1) Oxidation and deterioration of food ingredients caused by air oxidation and drying effect, resulting in rancidity, the loss of vitamins and linked browning. (2) Deterioration caused by microbial contamination and proliferation. (3) Decomposition of food caused by the endogenous enzymes such as oxidase, amylase and protease, generating thermal energy, water vapor and carbon dioxide gradually result in the deterioration of food. (4) Spoilage of food caused by the erosion and reproduction of insects, and directly or indirectly contaminated harmful substances.

5. (1) Rational use, harmless to human health; (2) No effect on the gastrointestinal flora; (3) degradable to the normal components of food in the digestive tract; (4) No effect on the use of antibiotics; (5) No harmful ingredients are produced during heat treatment.

Ⅳ. Translate the following words and expressions into Chinese.

酸性防腐剂　　　　　　　人工合成防腐剂
酶促褐变　　　　　　　　抑菌物质
杀菌作用　　　　　　　　抗菌谱
无机盐类防腐剂　　　　　复配防腐剂
山梨酸钾　　　　　　　　含硫化合物

Ⅴ. Translate the following expressions into English.

1. Acid preservatives mainly include benzoic acid, sorbic acid, propionic acid and their salts; the characteristic of acid preservatives is the greater acidity of the system, the better the preservative effect, but almost ineffective under alkaline conditions.

2. Natural food preservatives are a class of antiseptic substances separated and extracted directly from plants, animals, or microorganisms, also known as biological preservatives. According to different sources, natural food preservative can be divided into three categories: animal-derived, plant-derived and microbial-derived.

3. Food deterioration refers to the change process of the food quality (physical and chemical properties) under the influence of some certain factors (intrinsic and extrinsic).

LESSON 5 FOOD RISK ANALYSIS AND FOOD QUALITY

Ⅰ. Write true or false for the following statements according to the passage.

1. T 2. F 3. T 4. T 5. F 6. T 7. F 8. T 9. F 10. T

Ⅱ. Answer the following questions according to the passage.

1. Risk assessment is the use of existing scientific data to identify, confirm, and quantify the adverse consequences of exposure to hazardous substances in food.

2. The risk analysis assesses the various biological, physical, and chemical hazards affecting food

safety and quality, describes qualitatively or quantitatively the characteristics of risk, and proposes and implements risk management measures on the premise of referring to relevant factors, and conducts relevant risk assessments. Communication, which is the basis for formulating food safety standards.

3. Exposure assessment refers to a qualitative and/or quantitative assessment of the likelihood of biological, chemical, and physical factors ingested by food and, if relevant, exposure assessment from other sources.

4. Risk management is mainly divided into four parts: ① Preliminary risk management activities, including identifying food safety issues and clarifying their nature, describing risk conditions, determining risk management objectives, determining whether to conduct risk assessments and developing assessment policies, conducting risk assessments, and conducting analysis and classification of results; ② Determination of risk management options, including determining alternative management measures, evaluating alternative management measures and selecting optimal management measures; ③ Implementing management measures, including verifying the effectiveness of the necessary control systems and implementing control over selection measures and verification of implementation; ④ Monitoring and evaluating.

5. In the tangible product quality characteristics of foods, safety is the top priority. Secondly, food products have particularity in functionality and practicality. Thirdly, the comprehensive quality of food.

Ⅲ. Fill in the blanks according to the passage.

1. risk assessment, risk management, risk communication.

2. hazards, analysis

3. hazard identification, hazard characterization, exposure assessment, risk characterization.

4. functionality, credibility, security, adaptability

5. chemical substances, organisms, food-borne diseases

Ⅳ. Translate the following words and expressions into Chinese.

风险评估	食源性疾病
危害识别	生理调节性能
危害特征描述	风味嗜好性能
定量评估	消耗性产品
风险交流	国际食品法典委员会

Ⅴ. Translate the following expressions into English.

1. Hazards and analysis are two of the most basic concepts in food risk. It is important to clarify its definition for follow-up work. The CAC has a clear definition of the two. Hazard refers to the biological, chemical or physical factors or conditions in the food that may cause adverse health effects. Risk is the potential for adverse health effects and the severity of their effects, which leads to hazards in food.

2. In addition to the quality of tangible products, food quality includes process quality, service quality, and work quality. In food production, raw materials, production methods, production environment and other factors have a great influence on the quality of food. In addition, since food is a consumable product, the quality of its service is not reflected in the after-sales service, but reflected in the convenience of consumers' purchase and use.

LESSON 6 FOOD SAFETY TRACING

I. Write true or false for the following statements according to the passage.

1. T 2. T 3. F 4. T 5. F 6. T 7. F 8. F 9. T 10. T

II. Answer the following questions according to the passage.

1. Food geographical origin tracing is the guidelines, techniques, means and documents required by food companies to establish food trace and by the regulatory authorities in tracing, confirming and recalling.

2. Nowadays, a fingerprint map to distinguish the origin of agricultural products is going to be established based on the techniques including the mineral elements fingerprints analysis technology, electronic nose fingerprint technology, near infrared spectral analysis technology, DNA fingerprint techniques, metabonomics fingerprint technology combined with stoichiometry study, so as to trace back to the origin of different kinds of agricultural products.

3. In 2005.

4. DNA fingerprint is a kind of genetic marker based on the variation of nucleotide sequence among organisms that can directly detect the differences between individual organisms at DNA level, and it is a direct reflection of genetic variation at DNA level.

5. For example: how to screen the source traceability indicators effectively; the number of sample collection and the determination of the range of traceability; a variety of origin traceability technology fusion research; construction of an effective food source traceability data sharing platform; further development of corroboration technology; the cost of traceability technology and so on. With the unremitting research of agricultural researchers and the rapid development of agriculture in China, the technology will become mature and perfect and widely used and developed in all kinds of agricultural products and food.

III. Fill in the blanks according to the passage.

1. objectively, comprehensively, comprehensiveness

2. the mineral elements fingerprints analysis technology, near infrared spectral analysis technology, metabonomics fingerprint technology

3. The electronic nose

4. The source traceability technology of agricultural products

5. The genetic map, relative relationship

IV. Translate the following words and expressions into Chinese.

数理统计方法   挥发性成分
有效参考   多样性分析
聚类分析   传感器阵列
质谱   红外吸收光谱
电子鼻指纹图谱技术

V. Translate the following expressions into English.

1. Along with molecular biology, especially gene modification technology widely used in the field of crop breeding, there are a lot of similarity and genetic difference in morphology of new varieties, which

results in that the traditional morphological methods cannot be accurately identified.

2. DNA fingerprints traceability technology is mainly aimed at containing genes of genetic material, analyzing grains base sequence differences in gene, finally achieving the purpose of identification, distinguishing between grain varieties.

LESSON 7 FOOD SAFETY AND MICROBIAL INFLUENCE FACTORS

I. Write true or false for the following statements according to the passage.

1. T 2. F 3. T 4. T 5. F 6. F 7. T 8. T 9. T 10. F

II. Answer the following questions according to the passage.

1. These microorganisms in foods that cause human to get sick are termed food borne pathogens, including bacterial, fungal, viral, and parasitic (protozoa and worms) organisms.

2. Infectious food poisoning result from multiply live pathogenic organisms in the food ingestion.

3. Microbial contamination can occur during any of the steps in the farm-to-consumer continuum (production, harvest, processing, wholesale storage, transportation or retailing and handling family) and this contamination can arise from environmental (soil, water, feces, slurry), animal or human sources.

4. The extrinsic or environment related factors which affect the metabolism and multiplication of microorganisms include temperature, relative humidity, osmotic pressure, oxygen, inhibitors, light, and the type and quantity of microorganisms present in food.

5. The methods for controlling microorganisms are mainly heat, radiation, drying, ultrasonic, filtration, cleaning, microwave, high pressure, ohmic heating, far infrared, low temperature, application of chemical preservatives, biological preservatives (nisin, natamycin), bacteriophages, antagonistic bacteria, and a combination of antagonistic bacteria with bacteriophages.

III. Fill in the blanks according to the passage.

1. bacterial, fungal, viral, parasitic 2. Intoxications 3. endogenous, exogenous
4. intrinsic, extrinsic 5. salt, sugar, wine, vinegar, preservatives

IV. Translate the following words and expressions into Chinese.

食源性致病菌	单核细胞增生李斯特菌
枯草芽孢杆菌	金黄色葡萄球菌
拮抗细菌	大肠杆菌
小肠结肠炎耶尔森杆菌	相对湿度
肉毒梭状芽孢杆菌	渗透压

V. Translate the following expressions into English.

1. A variety of intrinsic and extrinsic factors affect the metabolism and multiplication of microorganisms. The intrinsic or food related factors are pH, moisture content, water activity, oxidation-reduction potential, nutrients, and the possible presence of natural antimicrobial agents. The extrinsic or environment related factors include temperature, relative humidity, osmotic pressure, oxygen, inhibitors, light, and the type and quantity of microorganisms present in food.

2. Physical methods can affect the chemical composition and metabolism of microbial growth, so the physical methods can be used to suppress or kill microorganisms. The physical methods for controlling microorganisms are mainly heating, radiation, drying, ultrasound, filtration, cleaning, microwave, high pressure, ohmic heating, far infrared, low temperature, and so on.

LESSON 8 FOOD QUALITY MANAGEMENT

Ⅰ. **Write true or false for each of the following statements based on what you have just read.**

1. T 2. F 3. F 4. F 5. F 6. T 7. T 8. T 9. F 10. T 11. T 12. F

Ⅱ. **Answer the following questions based on what you have just read.**

1. The principles of quality management arecustomer focus, leadership, employee engagement, process approach, improvement, evidence-based decision-making, and relationship management.

2. Good quality control requires that programs develop and maintain measurable indicators of quality, timely data collection and analysis, and effective supervision.

3. GAP are specific methods which, when applied to agriculture, create food for consumers or further processing that is safe and wholesome.

4. An individual SSOP should include: the equipment or affected area to be cleaned, identified by common name; the tools necessary to prepare the equipment or area to be cleaned; how to disassemble the area or equipment; the method of cleaning and sanitizing.

5. ISO standards help food producers meet legal and regulatory requirements, and give consumers the peace of mind that comes with knowing the food they consume meets high standards.

6. The food business operator (HACCP team) shall identify and register all potential (biological, chemical and physical) hazards that can have an adverse effect on the safety of the products.

7. For example, the ISO 9000 family addresses various aspects of quality management. The ISO 22000 family of International Standards addresses food safety management.

Ⅲ. **Fill in the blanks according to what you have just read.**

1. quality policy and strategy (QP&S), quality design (QD), quality control (QC), quality improvement (QI), quality assurance (QA)

2. Total quality management (TQM), quality cost analysis, strategy analysis

3. Quality improvement (QI), quality assurance (QA)

4. sanitary, manufacturing practices, HACCP

5. undesirable microorganisms, safe, adequate

6. produced, controlled

7. HACCP

8. safety hazards

Ⅳ. **Translate the following sentences into Chinese.**

1. 质量方针和策略（QP&S）是确保质量管理体系（QMS）包括在公司的长期业务战略中，并帮助公司采取适当的行动和分配资源以实现这些目标。

2. 当组织管理与其所有利益相关方的关系以其对绩效的产生最优化的影响时，更有可能实现持续的成功。

3. 土壤生物抗逆性

4. 验证程序

5. 危害分析

6. 关键控制点

7. 反作用

Ⅴ. Translate the following expressions into English.

1. Food Quality Management is a chain-oriented approach.

2. In many countries worldwide, legislation on the safety and suitability of foodstuffs requires hazard analysis and critical control point (HACCP) to be put in place by any food businesses or organizations, whether profit-making or not, whether public or private, carrying out any or all of the following activities: preparation, processing, manufacturing, packaging, storage, transportation, distribution, handling or offering for sale or supply of foodstuffs.

3. ISO 22000 is one of ISO's best-known standards. Within its broad scope, the ISO 22000 family includes standards specific to catering, food manufacturing, farming, packaging, and animal foodstuffs and feed production.

LESSON 9 QUALITY CONTROL TOOLS AND FOOD SAFETY EVALUATION

Ⅰ. Write true or false for the following statements.

1. T   2. T   3. F   4. T   5. T

Ⅱ. Multiple choice questions.

1. C   2. D   3. D   4. B   5. D   6. A   7. D   8. C   9. A   10. A

Ⅲ. Answer the following questions shortly.

①Quality control is a process that is used to ensure a certain level of quality in a product or service. It might include whatever actions a business deems necessary to provide for the control and verification of certain characteristics of a product or service. Most often, it involves thoroughly examining and testing the quality of products or the results of services.

②Food safety is a scientific discipline describing handling, preparation, and storage of food in ways that prevent food-borne illness. The occurrence of two or more cases of a similar illnesses resulting from the ingestion of a common food is known as a food-borne disease outbreak. This includes a number of routines that should be followed to avoid potential health hazards.

③It includes Hazard Analysis (HA) and Critical Control Points (CCP). The main works have seven parts. Conduct a hazard analysis, determine the CCPs, establish critical limits, establish a monitoring system, establish corrective actions, establish corrective actions, establish verification procedures, Establish documentation.

④Toxicity can simply be described as poisoning. This generally occurs when a substance or mixture of substances are sprayed or dusted onto animals, plants and those agents suffer negative effects afterward.

⑤Food borne disease refers to any illness resulting from the food spoilage of contaminated food, pathogenic bacteria, viruses, or parasites that contaminate food, as well as toxins such as poisonous mushrooms and various species of beans that have not been boiled for at least 10 minutes

IV. Translate the following words and expressions into Chinese.

1. 食品安全是一门用于防止食源性疾病，对食物进行有效处理和储存的学科。摄入普通食物导致两人以上患食源性疾病发生被称为食源性疾病暴发。考虑到食品是从市场到消费者，通常认为市场上的食品是安全的，并且能够为消费者安全送货和准备食品。

2. 为了减少人体对农药暴露，尽可能购买和食用有机认证食品。如果您所在地区有农贸市场，可以向他们询问农药的使用情况，许多小农选择不经过有机认证过程，但仍不使用农药。在食用前，不论食物是在哪里购买或种植的，都要彻底冲洗或清洗干净。在自己的花园中种植食物也是确保您拥有无农药食品的另一种方式。

V. Translate the following expressions into English.

1. HACCP ensures the safety of food in the process of consumption, production, processing, manufacturing, preparation and consumption, and is a scientific, rational and systematic method in hazard identification, evaluation and control. Identify the main risks that may occur in the food production process and take appropriate control measures to prevent the occurrence of hazards. By monitoring and controlling each step of the process, the probability of a hazard occurring is reduced.

2. Food safety hazards include factors that potentially damage or endanger food safety and quality, including biological, chemical, and physical hazards. Once food contains these hazards or is contaminated by these hazards, it can become a potentially hazardous food.

LESSON 10   CRISIS MANAGEMENT FOR FOOD INDUSTRY

I. Answer the following questions according to the passage.

1. A crisis is defined by the dictionary as a "critical moment or turning point".

A business book, define a crisis as a substantial, unforeseen circumstance that can potentially jeopardize a company's employees, customers, products, services, fiscal situation, or reputation. Both definitions contain an element of urgency that requires immediate decisions and actions from people involved.

ICM defines "crisis" as "a significant business disruption which stimulates extensive news media coverage. The resulting public scrutiny will affect the organization's normal operations and also could have a political, legal, financial, and governmental impact on its business".

2. A sudden crisis is defined as: a disruption in the company's business which occurs without warning and is likely to generate news coverage and may adversely impact:

    a. A business-related accident resulting in significant property damage that will disrupt normal business operations;

    b. The death or serious illness or injury of management, employees, contractors, customers, visitors, etc. as the result of a business-related accident;

    c. The sudden death or incapacitation of a key executive;

d. Discharge of hazardous chemicals or other materials into the environment.

3. The basic causes of a business crisis are four in number:

(1) Natural disaster (storms, earthquakes, volcanic action, etc.)

(2) Mechanical problems (ruptured pipes, metal fatigue, etc.)

(3) Human errors (the wrong valve was opened, miscommunication about what to do, etc.)

(4) Management decisions/indecision (the problem is not serious, nobody will find out)

4. A smoldering crisis is defined as: any serious business problem that is not generally known inside or outside the company, which may generate negative new-coverage if or when it goes "public" and could result in more than a predetermined amount in fines, penalties, legal damage awards, unbudgeted expenses and other costs. Examples of the type of smoldering business crises that would prompt a call to the crisis management team would include:

a. Sting operation by a news organization or government agency;

b. OSHA or EPA violations which could result in fines or legal action;

c. Customer allegations of overcharging or other improper conduct;

d. Investigation by a federal, state or local government agency;

e. Action by a disgruntled employee such as serious threat? or whistle-blowing;

f. Indications of significant legal/judicial/regulatory action against the business;

g. Discovery of serious internal problems that will have to be disclosed to employees, investors, customers, vendors and/or government officials.

5. Crisis management is the process of preparing for and responding to an unpredictable negative event to prevent it from escalating into an even bigger problem, or worse, exploding into a full-blown, widespread, life-threatening disaster. Crisis management involves the execution of well-coordinated actions to control the damage and preserve or restore public confidence in the system under crisis.

In the context of corporate governance, excellent crisis management is a "must" whenever a crisis occurs because of the crisis' enormous potential impact on the company's reputation and financial standing. Poor handling of a crisis situation can ruin the confidence of the customers or the public in a company and jeopardize its survival, a situation that normally takes a long time to correct, if it still is repairable at all. Such is the importance of public perception of a company's handling of a crisis situation that media coverage management has became an important ingredient of crisis management.

Crisis management doesn't start only when a crisis arises and ends when "the last fire has been put out". Crisis management requires actions before a crisis happens, while the crisis is unfolding, and after the crisis has ended.

6. There are many different ideas or theories on how to best manage a crisis situation. These differing ideas, nonetheless, have some common elements:

(1) The need to anticipate potential crisis situations and prepare for them;

(2) The need to provide accurate information during a crisis;

(3) The need to react as quickly as possible to the situation;

(4) The need for a response that comes from the top;

(5) The need for long-term solutions.

7. A comprehensive crisis management program includes the following components:

(1) An emergency response, which consists of all activities pertaining to safe management of immediate physical, health, and environmental effects of the crisis;

(2) Business continuity, or the company's ability to continue delivering goods and services despite the crisis;

(3) Crisis communications, which pertains to the internal and external PR management activities during a crisis;

(4) Humanitarian assistance, or the company's efforts to alleviate the physical, emotional, and psychological effects of the crisis on other people;

(5) Drills and exercises that allow personnel to rehearse what they need to do in a crisis situation

8. Within the food industry, the primary focus involves product quality issues related to product contamination, mislabeling, and product adulteration, which may result in the withdrawal/recall of products from the market place. The movement of food products between countries greatly increase the complexity.

Once a crisis situation is identified, it is critical to have a comprehensive crisis management plan to guide your company through the event. All functional groups within the company will be involved in the investigation, execution, or public relations of a recall. It is essential that each group and individual knows their role and has the resources available to accomplish their required tasks. Outside third party resources will often be required to support your efforts: public relations, media contacts, legal counsel, government liaisons, testing laboratories, university personnel, retrieval services, and crisis consulting services.

The crisis management plan must be an "evergreen" document, which is continuously maintained and updated to address the changing regulatory environment as well as internal personnel changes. "recall simulations" should be a part of the plan to regularly practice execution of the plan and to train personnel in real life situations. The regulatory requirements and the consumer environment vary greatly in different world areas. The U.S. plan will always need to be adapted for local conditions.

Ⅱ. **Choose a term from what we have learnt to fill in each of the following blanks. Change the word form where necessary.**

1. jeopardize, reputation
2. volcanic, rupture
3. sudden crisis, smoldering crisis
4. sudden smoldering
5. aftershocks
6. unpredictable negative
7. Pre-incident stage, Incident stage, Post-in cident stage
8. mislabeling, product adulteration